Springer Theses

Recognizing Outstanding Ph.D. Research

For further volumes:
http://www.springer.com/series/8790

Aims and Scope

The series "Springer Theses" brings together a selection of the very best Ph.D. theses from around the world and across the physical sciences. Nominated and endorsed by two recognized specialists, each published volume has been selected for its scientific excellence and the high impact of its contents for the pertinent field of research. For greater accessibility to non-specialists, the published versions include an extended introduction, as well as a foreword by the student's supervisor explaining the special relevance of the work for the field. As a whole, the series will provide a valuable resource both for newcomers to the research fields described, and for other scientists seeking detailed background information on special questions. Finally, it provides an accredited documentation of the valuable contributions made by today's younger generation of scientists.

Theses are accepted into the series by invited nomination only and must fulfill all of the following criteria

- They must be written in good English.
- The topic of should fall within the confines of Chemistry, Physics and related interdisciplinary fields such as Materials, Nanoscience, Chemical Engineering, Complex Systems and Biophysics.
- The work reported in the thesis must represent a significant scientific advance.
- If the thesis includes previously published material, permission to reproduce this must be gained from the respective copyright holder.
- They must have been examined and passed during the 12 months prior to nomination.
- Each thesis should include a foreword by the supervisor outlining the significance of its content.
- The theses should have a clearly defined structure including an introduction accessible to scientists not expert in that particular field.

Chiara Gualandi

Porous Polymeric Bioresorbable Scaffolds for Tissue Engineering

Doctoral Thesis accepted by
University of Bologna, Italy

 Springer

Author
Dr. Chiara Gualandi
Department of Chemistry
University of Bologna
Via Selmi 2
40126 Bologna
Italy
e-mail: c.gualandi@unibo.it

Supervisor
Prof. Mariastella Scandola
Department of Chemistry
University of Bologna
Via Selmi 2
40126 Bologna
Italy
e-mail: mariastella.scandola@unibo.it

ISSN 2190-5053   e-ISSN 2190-5061

ISBN 978-3-642-19271-5   e-ISBN 978-3-642-19272-2

DOI 10.1007/978-3-642-19272-2

Springer Heidelberg Dordrecht London New York

Cover design: eStudio Calamar, Berlin/Figueres

Printed on acid-free paper

Springer is part of Springer Science+Business Media (www.springer.com)

Parts of this thesis have been published in the following journal articles

Gualandi C., Wilczek P., Focarete M.L., Pasquinelli G., Kawalec M., Scandola M. "Bioresorbable electrospun nanofibrous scaffolds loaded with bioactive molecules" e-Polymer, publication no.: 060[2009] (Reproduced with Permission).

Zucchelli A., Fabiani D., Gualandi C., Focarete M.L. "An innovative and versatile approach to design highly porous, micropatterned, nanofibrous polymeric materials" Journal of Materials Science 2009, 44, 4969-4975 (Reproduced with Permission).

Focarete M.L., Gualandi C., Moroni L. "Working with electrospun scaffolds: some practical hints for tissue engineers" chapter 2, in "Electrospun Nanofibers Research: Recent Developments" A.K. Haghi Editor, Nova Science Publishers (Hauppauge, NY) (2009), p. 19-34 (Reproduced with Permission).

Gualandi C., White L. J., Chen L., Gross R. A., Shakesheff K. M., Howdle S. M., Scandola M. "Scaffold for tissue engineering from non-isothermal supercritical carbon dioxide foaming of a highly crystalline polyester" Acta Biomaterialia 2010, 6, 130-136 (Reproduced with Permission).

Foroni L., Dirani G., Gualandi C., Focarete M.L., Pasquinelli G. "Paraffin embedding allows effective analysis of proliferation, survival and immunophenotyping of cells cultured on poly(L-lactic acid) electrospun nanofiber scaffolds" Tissue Engineering part C: Methods 2010, 16(4), 751-760 (Reproduced with Permission).

Focarete M.L., Gualandi C., Scandola M., Govoni M., Giordano E., Foroni L., Valente S., Pasquinelli G., Gao W., Gross R.A. "Electrospun scaffolds of a polyhydroxyalkanoate consisting of ω-hydroxylpentadecanoate repeat units: fabrication and in vitro biocompatibility studies" Journal of Biomaterials Science, Polymer Edition 2010, 21, 1283-1296 (Reproduced with Permission).

Supervisor's Foreword

Tissue engineering, and regenerative medicine in general, are new challenges of modern medical science which aims to provide solutions to human society clinical needs. Since the beginning of these fields three decades ago, a huge amount of funding has been distributed worldwide to the scientific community with the final goal of creating an organ or part of it in vitro, or of inducing in vivo regeneration. Some of these research efforts have been awarded by seeing in vitro engineered tissues becoming commercial products that nowadays help people all around the world in completely or partially restoring the function of biological tissues. Despite the continuous expansion of tissue engineering (attested by an impressively growing number of publications and by the many academic institutions and industries working in the field), on the whole this area has been progressing slowly, probably due to the extremely difficult problems raised by the challenge of regenerating a biological tissue. Therefore, even if tissue engineering cannot be considered as a "new" discipline, the field needs innovative concepts and original approaches to overcome the present limits and face new challenges.

In the tissue engineering approach, biomaterials and scaffolds obtained therefore play a central role, i.e. they create the biophysical and biochemical environment suitable to direct cellular behavior and function. Success of an engineered tissue construct strongly depends on scaffold features and this is the reason why the main aim of this Ph.D. thesis was to develop innovative bioactive nanostructured constructs. The research activities, carried out not only at the Chemistry Department "G. Ciamician" (University of Bologna) but also at the Inorganic and Materials Chemistry Department and Centre of Biomolecular Science (University of Nottingham) and at the Polymers and Biomaterials Lab (University of Manchester), dealt with the manipulation of polymer chemical structure in order to tune material physical properties, and with optimization of scaffold 3D architecture by a smart use of different scaffold fabrication techniques. This thesis illustrates how a rational use of two alternative technologies, i.e. supercritical carbon dioxide foaming and electrospinning, can lead to fabrication of innovative polymeric bioresorbable scaffolds made of commercial or 'ad-hoc-synthesized' hydrolysable polyesters. The thesis also explores the possibility of making these new scaffolds bioactive by suitable functionalization procedures.

The multidisciplinary nature of this research is imperative in pursuing the challenge of tissue regeneration successfully. One of the strengths of this thesis is the integration of knowledge from chemistry, physics, engineering, materials science and biomedical science which has contributed to setting up new national and international collaborations, while strengthening existing ones.

Bologna, March 2011   Mariastella Scandola

Contents

List of Abbreviations

ATRP	Atom transfer radical polymerization
CE	2-Chloroethanol
CLF	Chloroform
C_p	Heat capacity
d	Needle to collector distance
DCM	Dichloromethane
DMEM	Dulbecco's modified eagle's medium
DMF	N,N-dimethylformamide
DMSO	Dimethyl sulfoxide
DSC	Differential scanning calorimetry
ECGS	Endothelial cell growth factor supplement
ECM	Extracellular matrix
ES	Electrospinning/electrospun
EtOH	Ethanol
FFPE	Formalin fixed paraffin embedding
GF	Growth factor
GMMA	Glycerol monomethacrylate
GPC	Gel permeation chromatography
HA	Hydroxyapatite
H_c	Crystallization enthalpy
HFIP	1,1,1,3,3,3-Hexafluoro-2-propanol
H_m	Melting enthalpy
μ-CT	Micro X-ray computed tomography
MetOH	Methanol
MI	ATRP Macroinitiator
M_n	Number average molecular weight
M_w	Weight average molecular weight
P(D,L)LA	Poly(D,L)lactide
P(L)LA	Poly(L)lactide
P(LA-TMC)	Poly((L)lactide-*co*-trimethylencarbonate)
P(PDL-CL)	Poly(ω-pentadecalactone-*co*-ε-caprolactone)

P(PDL-DO)	Poly(ω-pentadecalactone-*co*-p-dioxanone)
PBS	Phosphate buffered saline
PCL	Poly(ε-caprolactone)
PCR	Polymerase chain reduction
PDI	Polydispersity index
PEO	Poly(ethylene oxide)
PGA	Polyglycolide
PLA	Polylactide
PLA-T6	Six-arms star-branched oligo(D,L)lactic acid
PLA$_x$CL$_y$	Poly(lactide-*co*-ε-caprolactone) copolymers, where x and y indicate mol% content of the comonomers
PLA$_x$GA$_y$	Poly(lactide-*co*-glycolide) copolymers, where x and y indicate mol% content of the comonomers
PPDL	Poly(ω-pentadecalactone)
PTMC	Poly(1,3-trimethylene carbonate)
R	Solution flow rate
RH	Relative humidity
RT	Room temperature
ScCO$_2$	Supercritical carbon dioxide
SEM	Scanning electron microscopy
SI-ATRP	Surface initiated atom transfer radical polymerization
T_c	Crystallization temperature
TCP	β-Tricalcuim phosphate
TCPS	Tissue culture polystyrene
TE	Tissue engineering
T_g	Glass transition temperature
TGA	Thermogravimetric analysis
THF	Tetrahydrofuran
T_m	Melting temperature
WAXD	Wide angle X-ray diffraction
ΔV	Applied voltage

Chapter 1
Introduction

1.1 Tissue Engineering

Tissue or organ failure is one of the most frequent and devastating problems in medicine. Current therapeutic approaches (allografts, xenografts, autografts and implantation of biomedical devices) have largely improved the quality of life but they are associated with clear limitations including donor availability, infection, poor integration and potential rejection of the implant. Regenerative medicine was developed with the aim to overcome such limitations and to find revolutionary and powerful therapies for the treatment of tissue diseases, with the basic idea to repair or re-create tissues or organs in order to restore impaired functions [1, 2].

Since its origin, regenerative medicine has rapidly grown and has attracted the interest of many scientists and surgeons throughout the world. Nowadays regenerative medicine encompasses different strategies for the creation of new tissue including the use of cloning, of isolated cells, of non-cellular structures and of cells constructs. The latter approach, which is usually referred to as tissue engineering (TE), is believed to be highly promising for regenerating tissues. It is pointed out that a clear distinction between TE and regenerative medicine does not exist in the literature and some scientists use these terms as synonyms. In the present project, however, TE will be considered a sub-discipline of regenerative medicine that intends to use cell-constructs to achieve tissue repair.

Even if the term "tissue engineering" was firstly coined in the mid of 1980s, it became part of the scientists common language only in 1993, when Langer and Vacanti [3] defined TE as "an interdisciplinary field that applies the principles of engineering and life sciences toward the development of biological substitutes that restore, maintain, or improve tissue function or a whole organ". Figure 1.1 sketches TE approach for the preparation of cell constructs: cells are cultured in vitro under precisely controlled culture conditions on a porous three-dimensional (3D) material that acts as scaffold for cell growth and proliferation. Once being

C. Gualandi, *Porous Polymeric Bioresorbable Scaffolds for Tissue Engineering*,
Springer Theses, DOI: 10.1007/978-3-642-19272-2_1,
© Springer-Verlag Berlin Heidelberg 2011

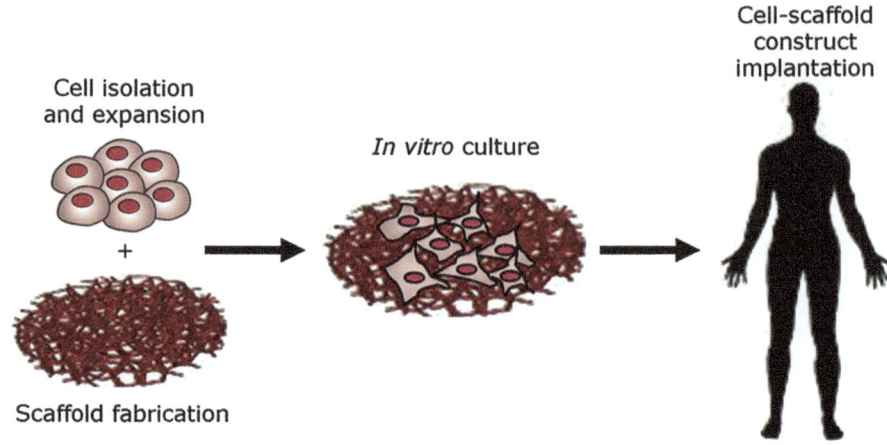

Fig. 1.1 TE approach by using cell-scaffold construct

implanted in the human body, the scaffold is eventually bioresorbed and the space occupied by it is replaced by new tissue produced by cells [4, 5].

Significant progress has been realized in TE since its principles were defined and to date several products, incorporating cells together with scaffold, have gained regulatory approval. Commercial devices are mostly dedicated to skin, bone and cartilage engineering whereas products for cardiac, nerve, kidney or pancreas engineering are not widespread yet [6].

The obtainment of an engineered tissue encompasses the optimization of a huge number of factors and parameters that interplay in affecting the success or the failure of the construct. Since an engineered tissue is basically composed of cells seeded on a scaffold, the design of a cell-scaffold construct firstly requires the selection of both suitable cell type and porous scaffold, depending on the characteristics of the tissue that has to be replaced.

In terms of cell origin, cells can be autologous (patient's own cells harvested by biopsy), allogeneic (cells from other human sources) or xenogeneic (cells from different species), although the last two, being genetically different from patient's cells, may induce immunorejective response. Moreover, cells exist in a differentiate state or they can be undifferentiated (stem cells). In the first case cell phenotype and functions are well defined but this is accompanied with a limited capability to proliferate. On the contrary stem cells can self-renew for long periods of time and can be induced to differentiate upon exposure to specific cues [7]. The rapid expansion of interest in stem cells arises from their proliferative and developmental potential that promises an essentially unlimited supply of specific cell types for both basic research and transplantation therapies.

The second basic component of an engineered tissue is the scaffold that acts as a temporary template, guiding cell organization, growth and differentiation and providing structural stability and a 3D environment where cells can produce new biological tissue. Scaffolds may be created from various types of materials, including natural and synthetic polymers and inorganic substances. The fabrication technique controls the morphology of the porous scaffold that can be made of fibres, gels or can be obtained by creating pores within a polymer matrix.

Additional components of cell-scaffold constructs can be biomolecules that are introduced either as scaffold additives to be delivered during cell culture or by functionalizing scaffold surface. Biomolecules (e.g. growth factors, peptide sequences, gene vectors, etc.) are used in order to induce specific cell behaviour (e.g. enhancing cell adhesion, cell proliferation, cell differentiation, etc.).

Another important aspect in the fabrication of an engineered tissue are cell culture conditions that are known to influence cell growth. In particular, dynamic cell culture conditions improve the development of a more tissue-like construct with respect to static conditions, thanks to efficient nutrition and exogenous stimuli that direct cellular activity and phenotype [8–10]. Dynamic cell cultures are performed with bioreactors, where culture conditions, such as mass transport throughout the scaffold, mechanical stimulation, electrical stimulation etc., can be precisely controlled.

In summary, several critical elements should be considered in order to achieve successful regeneration of damaged tissue, including cell type and source, scaffold material and 3D structure, biomolecules and cell culture conditions (Fig. 1.2). Hence, it is clear that TE studies require a multidisciplinary approach that must involve not only biological and medical expertises but also competences in engineering, chemistry and materials science.

The present research project focuses its attention on the fabrication of polymeric bioresorbable scaffolds by means of different techniques, with a particular consideration towards electrospinning (ES) technology. Knowledge of polymer science has been applied to produce, manipulate and characterize polymeric scaffolds. Given the multidisciplinary character of this field, the research was carried out with the cooperation of experts in other disciplines other than polymer science, such as mechanical and electrical engineering and biochemistry. In particular, engineers contributed to improve scaffold fabrication technology whereas biochemists

Fig. 1.2 TE key elements

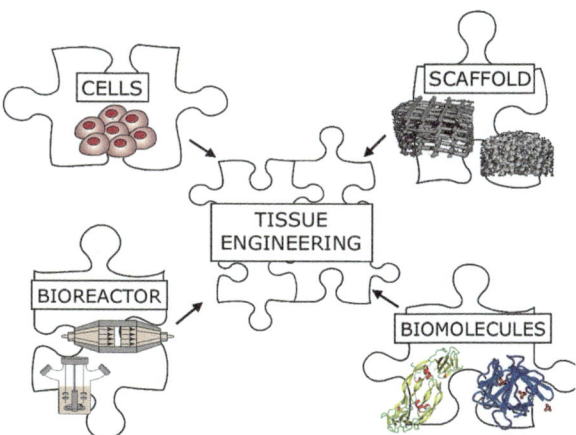

performed in vitro biological experiments making use of the polymeric scaffolds produced in the context of this project.

1.2 Biological Tissues

A detailed understanding of the structure and function of normal biological tissues is central to the design of artificial organs and to the development of tissue engineering strategies. In general, biological tissues are composed of cells and of a non-cellular part called extracellular matrix (ECM). Physiological functions in the human body are coordinated by different types of organs composed of a great variety of mammalian cells surrounded by ECM, which displays different chemical compositions and different spatially organized configurations depending on the type of organ.

Cells are tissue building blocks and a considerable part of TE studies deal with their isolation from native tissues and their expansion in vitro. Since most cells do not grow efficiently in suspension but they need to be attached to a substratum [11], the common procedure is to seed cells on artificial bidimensional substrates called tissue culture polystyrene (TCPS) plates. Once enough cells are obtained, they are harvested from the 2D plates and then seeded on 3D scaffolds. In conventional TE approaches fully differentiated adult cells, which are estimated to be more than 210 types in mammalian body [7], are employed. However, given the shortage of adult human cells due, in particular, to difficult harvesting procedure and their limited survival capability in vitro, modern TE approaches involve the use of stem cells, i.e. of undifferentiated cells that easily duplicate in vitro and that can be induced to specialize following exposure to specific cues [2].

Cell behaviour is deeply influenced by hierarchical patterns and by physico-chemical properties of the environment (Fig. 1.3). In all biological tissues, cells are surrounded by ECM that is composed of carbohydrates and proteins locally secreted by cells and assembled in an organized network. For the sake of simplicity ECM can be considered composed of (1) a gel-like component, (2) a fibrous component and (3) specialized proteins [12]. The highly viscous and hydrated gel-like part confers lubricant and hit-absorption properties to the ECM. It consists of carbohydrates assembled to form polysaccharides commonly called glycosaminoglycans which, in turn, are covalently attached to a protein backbone to form proteoglycans. The fibrous component of ECM is essentially composed of collagen and elastin that create a complex network of fibres with diameters ranging from few to hundreds of nanometers, imparting rigidity and strength to the entire tissue. Finally, ECM holds proteins such as growth factors (GFs), cytokines, enzymes and multidomain proteins (e.g. fibronectin, laminin, vitronectin, etc.) that play a key function in the communication with the surrounding cells.

Besides providing structural support to cells, ECM plays a central role in modulating cell behaviour and in maintaining tissue architecture and functions thanks to its dynamic interaction with cells. Indeed, cells continuously interact with the external environment via membrane proteins (receptors, e.g. integrins)

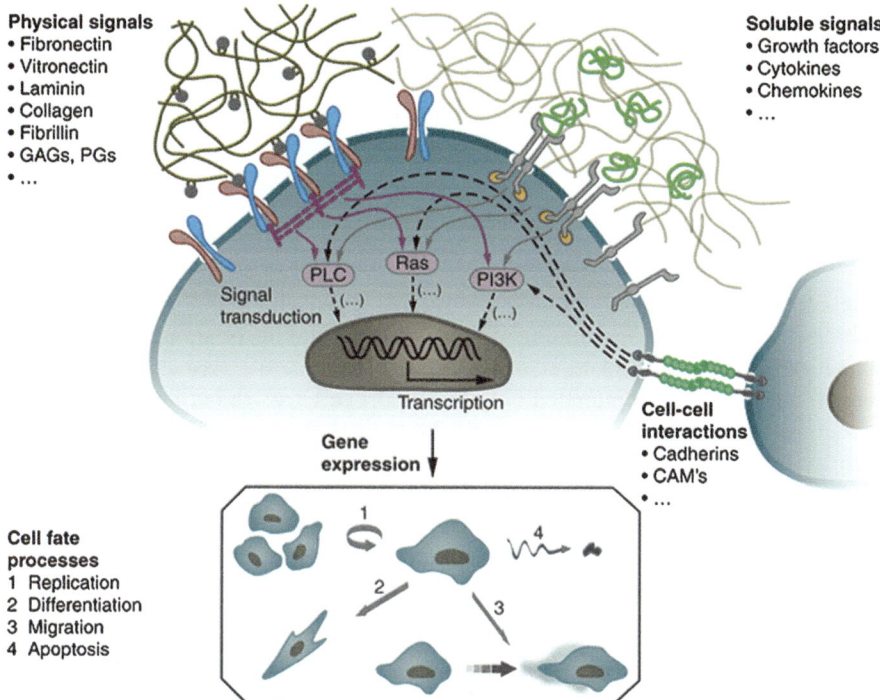

Fig. 1.3 Schematic representation of the reciprocal molecular interaction between cell and its surrounding (reprinted from [12], Copyright (2005), with permission from Macmillan Publishers Ltd: Nature Biotechnology)

that bind external proteins (ligands, e.g. fibronectin) located both on the surface of surrounding cells and in the ECM, by following a lock-and-key mechanism. Through an intracellular cascade of reactions, the ligand-receptor interaction is translated into a specific signal to guide the cell to a specific activity [13]. This dynamic interaction allows to finely control cell fate, shape and behaviour in response to even small changes in ECM composition [14–17]. Overall, ECM functions can be summarized as follows [18]:

- establishment of a hierarchical patterned micro/nanoenvironment;
- mechanical and structural support;
- regulation of cell shape and cell polarity;
- storage of regulatory molecules (enzymes, GFs, multidomain proteins);
- regulation of cell function (e.g. proliferation, growth, survival, migration, and differentiation).

Given the central role of ECM and of cell environment in determining cell response and behaviour, it is quite evident that biologists and biochemists need to deeply understand the biological phenomena that rule cell-ECM and cell–cell interactions. It is also clear that cells must be provided with a scaffold, having suitable biological and mechanical features to ensure cell attachment, proliferation and spontaneous deposition of ECM by cells. To this aim, materials science plays its role in TE field, by fabricating an ECM-substitute scaffold with appropriate physical and chemical properties and with a proper 3D structure and architecture.

1.3 The Scaffold

The officially accepted definition considers a scaffold as a "support, delivery vehicle or matrix for facilitating the migration, binding or transport of cells or bioactive molecules used to replace, repair or regenerate tissues" [19]. Indeed, recent scaffolds are not intended only to support cell growth but they can also load bioactive molecules having a specific biological function. This aspect will be deeply discussed in Sect. 1.4. Moreover, it is worth noting that last generation scaffolds are intended to be as much biomimetic as possible, in terms of (1) mechanical performances, (2) 3D morphology and (3) surface chemistry, and they are designed according to this scope.

It is well-accepted that an ideal functional scaffold should meet the following challenging requirements [20–23]:

- to be biocompatible;
- to have mechanical properties consistent with those of the tissue it replaces;
- to be bioresorbable;
- to degrade at a rate matching that of new tissue formation;
- to have proper surface properties to enable cell attachment, growth, proliferation, and differentiation as well as to promote extracellular matrix formation;
- to have the optimum architectural properties in terms of pore size, porosity, pore interconnectivity, and permeability in order to allow efficiently delivery of nutrients and removal of waste.

The imperative requirement for a scaffold is to be biocompatible. Since the scaffold works in contact with leaving cells in vitro and with tissue, once implanted in vivo, it must not elicit harmful response from the biological environment, i.e. it should interact with cells and host tissue without inducing cytotoxicity or adverse immune response.

Scaffold mechanical properties are another key element that distinguishes a successful implant from a failed one. Indeed, since many tissues undergo mechanical stresses and strains, it is extremely important that mechanical properties of the scaffold match as closely as possible those of the tissue intended to regenerate, so that formation of new ECM is not limited by mechanical failure of the scaffold. Moreover, a good mechanical transfer between the scaffold and the new forming tissue is required in order to provide sufficient mechanical stimulation for ensuring a proper tissue growth [24].

Scaffold is intended to be a temporary support that is eventually replaced in the organism by new regenerated tissue. Hence, scaffold must be degradable in the human body through molecular fragmentation mechanisms that results in the formation of degradation by-products and the gradual disappearance of the scaffold. Degradation can occur via hydrolysis or it can be mediated by enzymes, depending on polymer chemical structure. In any case, the long-term success of an engineered tissue will be achieved only if degradation products are completely resorbed by the organism by naturally occurring metabolic pathways, i.e. if the scaffold is bioresorbable [25]. Moreover, scaffold degradation rate should mirror the rate of tissue formation. This criterion is extremely difficult to achieve but it is particularly important as regards the structural supporting role of the scaffold. Indeed, in order not to compromise the integrity of the implant, the scaffold should maintain its structural function until the regenerated tissue can assume its supporting role and, over time, the degrading scaffold should gradually transfer its function of load bearing to the new forming tissue. The control of scaffold mechanical properties over time and during degradation process remains one of the greatest challenges in tissue engineering [26, 27].

Another significant feature of the scaffold is its surface properties which are directly connected to its capability of creating a chemically suitable environment that promotes cell adhesion, migration and proliferation in order to obtain an entirely colonized 3D cell construct. Cell adhesion is always a receptor-mediated process that occurs via interaction between membrane proteins, called integrins, and multidomain proteins that act as ligands. In natural tissues typical ligands dislocated in the ECM are fibronectin, laminin and vitronectin that bind integrins, forcing the cells to attach to the ECM [28, 29]. Cell-scaffold interaction similarly occurs. Indeed, when a scaffold gets in contact with a biological environment the first occurring event is protein absorption as a result of Van der Waals, hydrophobic and electrostatic interactions, and hydrogen bonding [30]. These interactions depend on scaffold surface, namely on its chemical composition, roughness and topography [31]. Therefore cell-scaffold interactions are related to the composition of the protein layer attached to the scaffold surface. It has been largely demonstrated that polymers can adsorb many proteins [32] and it is also possible to adapt a material presenting good bulk properties by improving its surface properties towards cell adhesion through surface modification. Many techniques have been developed to modify material surfaces such as plasma or ion treatment but, more recently, ECM biomolecules, such as proteins, peptides or growth factors, have been immobilized on scaffold surface with the aim to obtain bioactive and biomimetic scaffolds [33]. Surface modification approaches will be more extensively discussed in Sect. 1.4.

Cell colonization of the scaffold depends not only on scaffold surface properties but also, indirectly, on its 3D architecture. Porosity (the amount of void space), size, geometry, orientation and interconnectivity of pores and channels directly affect the transport and delivery of nutrients for cells throughout the scaffold [34]. In particular high porosity, high surface area to volume ratio and high pore interconnectivity are required in order to ensure uniform tissue ingrowth, efficient delivery of nutrients to the interior of the scaffold and removal of waste towards the exterior [35]. The supply of nutrients and oxygen is realized in vivo by the blood vascular system. The achievement of vascularization of 3D scaffolds is still one of the greatest challenges in TE.

The above described scaffold characteristics (i.e. biocompatibility, mechanical properties, bioresorbability, surface properties and architecture) are strictly related to two major factors that interplay in controlling scaffold properties: (1) the type of polymer material and (2) the scaffold fabrication technology. For instance, toxic residual substances can be released out from the scaffold causing harmful effects. Such substances can be either monomers or impurities in the starting material or substances deriving from material processing (e.g. degradation products, organic solvents, etc.) Mechanical properties primary depend on the raw material but they can dramatically change according to scaffold architecture which, in turn, is determined by the technique employed to produce the scaffold. Another example is bioresorbability and, in particular, the degradation rate that depends not only on polymer properties (i.e. chemical composition, monomer distribution and micro-structure in copolymers and molecular weight) but also on scaffold architecture (i.e. scaffold dimension and pore walls).

It is clear that the correct design of a scaffold for a specific application requires to accurately know which properties it should exhibit in order to successfully achieve its function. Once the necessary scaffold features are clearly defined, material science intervenes in selecting the proper polymer material, the suitable scaffold fabrication technology and the appropriate treatments in order to obtain a scaffold that matches as closely as possible the specific requirements.

The following subchapters outline the principles of biomaterials science and describe the more frequently used biomaterials for tissue engineering applications, with particular attention towards polymeric ones. Common scaffold fabrication approaches are introduced, focusing mainly on electrospinning and supercritical carbon dioxide technology.

1.3.1 Biomaterials

A biomaterial can be considered a synthetic material used to make a device designed to replace a part or a function of the body. The commonly accepted definition of biomaterial was proposed at the Conference of the European Society for Biomaterials (England, 1986): "any substance, other than a drug, or combi-nation of substances, synthetic or natural in origin, which can be used for any period of time, as a whole or as a part of a system which treats, augments, or replaces any tissue, organ, or function of the body" [36].

The basic requirement of a biomaterial is to be biocompatible, namely "to perform with an appropriate host response in a specific application" [36]. The biomaterial itself, but also additives or degradation products, must not cause harmful reactions in contact with the body. For this reason the design and the fabrication of a biomedical device, in terms of synthesis of the raw materials but also of manufacture technologies, must respect, first of all, the biocompatibility requirement. Biocompatibility must be tested and the devise must be approved by appropriate regulatory agencies (such as the Food and Drug Administration, USA) before it can be marketed.

According to the definition of biomaterial reported above, the following devices fall within this category:

- implantable devices, e.g. dental implants, pacemakers, orthopedic and vascular prostheses, etc.;
- devices working in contact with biological tissues/fluids for a limited period of time, e.g. surgical instruments, catheters, contact lenses, sutures, etc.;
- devices for extra-body treatments, e.g. dialysis membranes, blood vessels.

According to their chemical nature, synthetic biomaterials can be classified as: (1) polymers, (2) metals, (3) ceramics and (4) composites. Table 1.1 reports some examples of applications, together with some considerations about advantages and disadvantages related to each class of biomaterial.

Thanks to the continuous progress in molecular biology and in the comprehension of cell-biomaterial interactions, since its birth, biomaterial science has seen an extraordinary evolution in the development of increasingly biocompatible, bioactive and specifically functional biomedical devices. The improvement of biomaterial features went through three different stages, each concerning different purposes [38]. The "first generation" of biomaterials, developed in 1960s as implantable devices, were designed to possess suitable physical properties in view of their function as organ substitutes, and they were intended to be inert, namely to

Table 1.1 Categories of biomaterials (adapted from [37])

Materials	Uses	Advantages	Disadvantages
Polymers			
Polyammides, silicon rubber, polyesters, polyurethans, polytetrafluoroethylene, polymethylmetharylate, etc.	Sutures, blood vessels, catheters, devices for drug delivery, contact lens, etc.	Resilient Easy to fabricate Wide range of mechanical properties Biodegradable	Not strong Deformation with time
Metals			
Ti and its alloys, stainless steels, Au, Ag, Co–Cr alloys, etc.	Joint replacements, bone plates and screws, dental implants, surgical instruments, etc.	Strong Tough Ductile	May corrode High densities
Ceramics			
Aluminium oxides, calcium phosphates, hydroxyapatite, Bioglasses, etc.	Dental implants, femoral heads, coating of orthopaedic devices	Strong under compression Good trybological properties	Brittle Not resilient
Composites			
	Joint implants, tendon and hip replacements, heart valves	Strong Tailor-made	Difficult to fabricate

cause minimal tissue reactions. Later in the 1980s, the "second generation" of biomaterials was born with the necessity not only to be tolerated by the organism, but also to elicit desired and controlled response by the biological tissue, again with no harmful effects. The so called bioactive materials were used, for example, in orthopaedic surgery (Bioglass®). Moreover, in this period problems connected with long term tissue-biomaterial interactions were minimized with the development of bioresorbable materials that were used also as drug delivery systems. Subsequently biomaterials have evolved into their "third generation" and nowadays they are intended to stimulate highly precise reactions at the molecular level with biological tissues. With this aim it is clear that precise surface engineering and nanotechnology are needed in order to build tailored architectures for specific applications. Scaffolds for tissue engineering can be considered as "third generation" biomaterials.

For this reason it is clear that raw biomaterials suitable to produce bioresorbable scaffolds are exclusively either biodegradable polymers (either natural or synthetic ones) or biodegradable ceramics (calcium phosphates) that must disappear in the human body as a consequence of hydrolytic and/or enzymatic degradation, whereas metals and all the other non-degradable materials are not used in this context.

Biodegradable ceramics, most of all hydroxyapatite (HA) and β-tricalcium phosphate (β-TCP), are naturally found in bone and they are used in combination with degradable polymers to make composite scaffolds for bone tissue regeneration [39–41]. Indeed, these inorganic substances are often difficult to be processed by their own into highly porous structures. Moreover, their inherent brittleness limits their use as plain materials for scaffold fabrication, so that, in order to obtain structural resistant supports, it is necessary to combine them with polymer matrices [21, 42].

Natural polymers such as proteins (e.g. collagen, gelatin, fibrin, starch, etc.), polysaccharides (e.g. hyaluronic acid, alginate, chitin, etc.) and bacterial polyesters (e.g. polyhydroxybutyrate, polyhydroxyvalerate, etc.) are widely used to produce scaffolds with different topography, either as pure materials or in combination with synthetic polymers or inorganic substances [43–47]. The use of natural polymers ensures excellent bioactivity towards biological environment because of their inherent properties of biological recognition. In particular, if ECM polymers, such as collagen or hyaluronic acid, are employed, scaffolds that mimic the chemical properties of natural ECM are obtained (biomimicry). Biomimetic features are so relevant that, besides pure natural polymers, decellularized ECM, containing multiple natural macromolecules, is also used as scaffold for tissue repair. Despite the incredible advantage of biomimicry, some issues, associated with purification, pathogen transmission, processability into porous structure and sustainable production, restrict the use of natural polymers for scaffold fabrication. Moreover, poor control of mechanical properties and degradation rate limits the possibility to tailor their employment to specific functions.

Some problems associated with the use of natural materials can be overcome with the employment of synthetic polymers. Indeed, the latter, besides being less expensive and better processable, can be synthesized ad hoc with a precise control of molecular structure to tune mechanical and degradation properties. Furthermore, polymers can be synthesized with specific functional groups in order to make them bioactive towards biological environment. These characteristics make synthetic polymers extremely attractive raw materials for scaffold fabrication. The molecular structures of the most common ones [21, 48, 49] are reported in Fig. 1.4.

Scientific literature focuses its efforts mainly in the development of scaffolds made of polyesters, which possess the most suitable features to be employed in this context. This class of materials can degrade in water as a consequence of ester bond hydrolysis. It is pointed out, however, that hydrolysis occurs depending mainly on material hydrophilicity and, for this reason, not all polyesters can be considered as hydrolizable materials (the aromatic polyester polyethylenterephtalate, for example, is highly stable in aqueous environment). It is worth noting that one potential concern arises from the local pH changes that can occurs during degradation as a consequence of acid nature of degradation products [50].

Currently, the most widely investigated and most commonly used biomedical polyesters are polylactide (PLA), polyglycolide (PGA) and their copolymers (PLA_xGA_y, where x and y indicate mol% content of the monomers) [51], usually referred to as poly-α-hydroxyacids (Fig. 1.4). These materials degrade upon water exposure into products absorbable by the organism: lactic acid, that is normally produced by muscular contraction, can be metabolized through the citric acid cycle whereas glycolic acid may be eliminated directly in urine or may be converted to

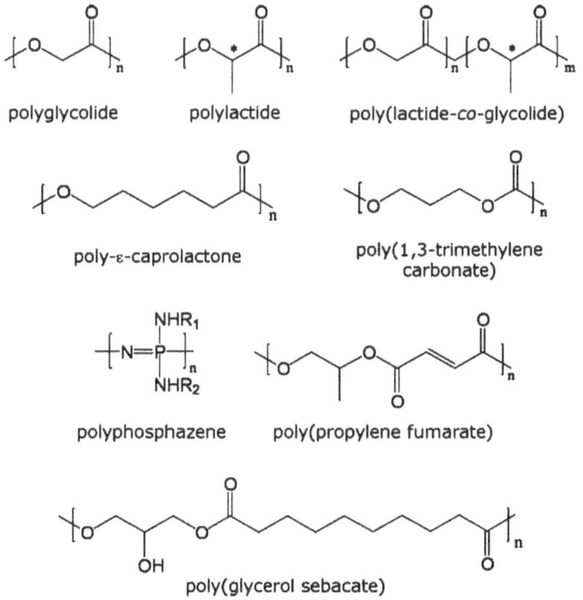

Fig. 1.4 Common synthetic biodegradable polymers for scaffold-based TE

polyglycolide   polylactide   poly(lactide-co-glycolide)

poly-ε-caprolactone   poly(1,3-trimethylene carbonate)

polyphosphazene   poly(propylene fumarate)

poly(glycerol sebacate)

enter the citric acid cycle [52]. Poly-α-hydroxyacids are usually synthesized by ring-opening polymerization of the cyclic dimers of lactic acid and glycolic acid (lactide and glycolide). Lactic acid is a chiral molecule existing as L or D isomer, thus polylactide can be optically pure, poly(L)lactide or poly(D)lactide, or it can exist in the racemic form, poly(D,L)lactide, depending on the starting monomer. PGA and the stereoregular forms of PLA are semicrystalline polymers whereas PLAGA copolymers can be amorphous depending on molecular composition because the presence of a comonomer disturbs the crystallization ability of the chains. It has been reported that PLAGA random copolymers with L-lactic acid units are amorphous when glycolic acid amount falls within the range 25–75%, whereas PLAGA copolymers with lactic acid in both L and D forms are amorphous in the range 0–75% of glycolic acid content [53]. Copolymerization modulates not only the phase morphology (amorphous to crystalline phase ratio) and the thermal properties (glass transition temperature, crystallization and melting temperature) but also the degradation rate. Degradation mechanism and kinetics of this series of polymers have been intensively studied [54, 55]. Hydrolysis rate is primarily correlated with hydrophobicity of the material that changes depending on the copolymer composition, being lactic acid more hydrophobic than glycolic acid because of the presence of an extra CH_3 group. The hydrophobicity of lactic acid limits the water uptake and, accordingly, the homopolymer PGA should be the faster degrading one, with degradation rate that should decrease with the increase of lactic acid content. However, hydrophobicity is not the only parameter affecting degradation rate, which is deeply influenced also by phase morphology. Indeed, molecular chains packed in the crystalline phase absorb a lower amount of water with respect to the less dense amorphous phase. This fact explains why the amorphous $PLA_{50}GA_{50}$ degrades faster than the semicrystalline PGA, even if the latter is more hydrophilic [56, 57]. Other factors affecting degradation kinetics are polymer molecular weight and shape of the device. The possibility to modulate, within a wide range, degradation rate and mechanical properties by controlling copolymer composition and molecular structure makes these polyesters the most widely employed bioresorbable materials suitable for many and different applications, such as resorbable surgical sutures, drug delivery systems, orthopaedic appliances and scaffolds.

Polyesters with a long carbon backbone chain are identified as ω-polyhydroxyalcanoates among which poly(ε-caprolactone) (PCL) is frequently employed to produce scaffolds (Fig. 1.4). Being these polymers highly hydrophobic and semicrystalline, their degradation kinetic is extremely slow, thus they are suitable for long-term tissue engineering applications. In particular, ε-caprolactone is largely used also as comonomer to slow down degradation rate of homopolymers such as PLA (giving rise to PLA_xCL_y copolymers) [58–60].

Aliphatic polycarbonates are also applied in tissue engineering [61], in particular poly(1,3-trimethylene carbonate) (PTMC), an elastomeric polymer potentially candidate for soft tissue engineering. Trimethylene carbonate is also widely used as comonomer with lactic acid, glicolic acid or ε-caprolactone in order to obtain scaffolds with elastomeric properties [62–66].

Other categories of materials are currently under investigation [67]:

- *Polyphosphazenes* these polymers with a backbone of alternating phosphorus and nitrogen atoms are at the interface between inorganic and organic polymers. Biodegradable polyphosphazenes can be synthesized by incorporating side groups on phosphorous atoms with the possibility to modulate the degradation rate over hours, days, months, or years by carefully controlling the nature and composition of side substitutes [68, 69]. Thanks to their synthetic flexibility, good biocompatibility, non-toxic degradation products and tailored mechanical properties they are good candidates for various soft and hard tissue engineering applications [70–72].
- *Poly(propylene fumarate)* being available as an injectable system that is cross-linked in situ, it is a very interesting material for bone tissue application in the treatment of crevices and defects. Its mechanical properties vary according to the cross-linking agents used and they are often improved by the addiction of inorganic particles such as β-TCP [73–75].
- *Poly(glycerol sebacate)* it is a biodegradable and biocompatible material with very interesting elastomeric properties that make it a suitable polymer for soft tissue engineering [76–79].

1.3.2 Scaffold Fabrication Technologies

Selection of the raw material for producing the scaffold is complementary to the choice of the proper fabrication technology suitable to achieve the desired scaffold properties. Since TE developed, many techniques have been used to this aim and even nowadays new methods are invented, though they are mostly modification or smart combination of already existing techniques. This section aims at providing an overview of conventional and well-known processes for scaffold fabrication by illustrating the characteristics of the obtained scaffold [20, 21, 23, 26, 80, 81]. Figure 1.5 reports representative images of scaffolds obtained by the techniques described in this section.

1.3.2.1 Textile Technologies

Earlier TE scaffolds composed of fibrous biodegradable polymer fabrics were produced using textile technologies. Mostly PGA and PLA non-woven scaffolds were used in TE research [89–91]. These fibrous structures have high porosity (up to 95%) with interconnected pores and high surface area to volume ratio, though fibre diameter is confined in the range 10–15 μm.

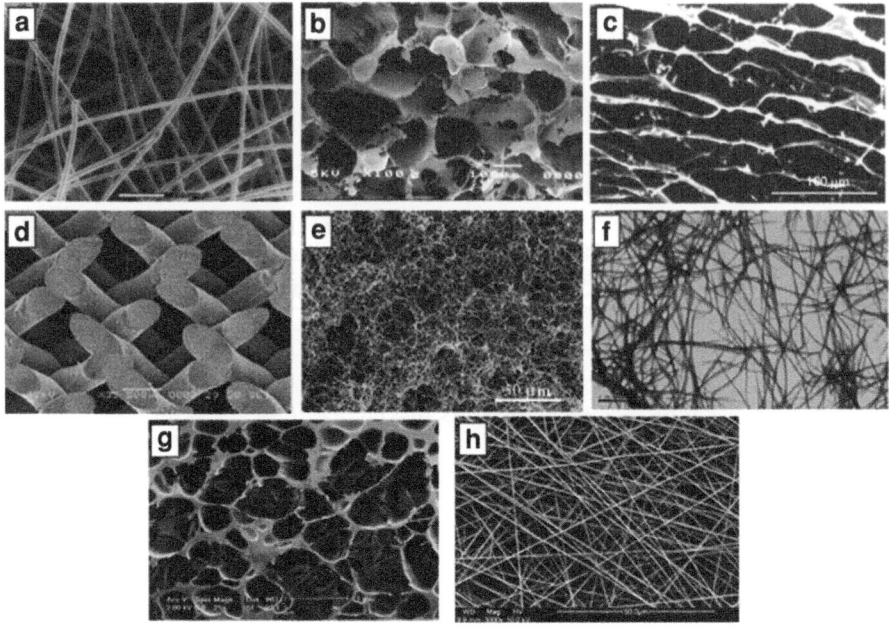

Fig. 1.5 Scanning Electron Microscope images of representative scaffolds obtained by different fabrication technologies: **a** textile technologies (reprinted from [21], Copyright (2004), with permission from Elsevier), **b** solvent casting and particulate leaching (reprinted from [82], Copyright (2004), with permission from Elsevier), **c** freeze drying (reprinted from [83], Copyright (1999), with permission from Elsevier), **d** solid freeform fabrication (reprinted from [84], Copyright (2003), with permission from Elsevier), **e** thermally induced phase separation (reprinted from [85], Copyright (1999), with permission from Wiley), **f** peptide assembly (reprinted from [86], Copyright (2006), with permission from Elsevier, transmission electron microscopy image), **g** gas foaming (reprinted from [87], Copyright (2005), with permission from Wiley), and **h** electrospinning (reprinted from [88], Copyright (2008), with permission from Elsevier)

1.3.2.2 Solvent Casting and Particulate Leaching

This technique, firstly described by Mikos et al. in 1994 [92], uses a water-soluble porogen to produce pores within a polymer matrix. In brief, a polymer solution is cast in a mould containing particles of desired dimension. After solvent evaporation, particles are leached out by immersion in water. Mainly PLA and PLAGA scaffolds were produced with this approach. Salts are the most commonly used porogen [64, 92], but also sugar [93], paraffin [94] and gelatine spheres [94] have been employed. This method enables to tune independently pore size (up to 500 μm in diameter) and porosity (up to 90%) by controlling particle dimensions and porogen/polymer ratio respectively. However, due to gravity, homogeneous pore distribution is only obtained in scaffolds less than a few millimetres thick [95] and complete elimination of both organic solvent and porogen is rather difficult to achieve if pores are not completely interconnected. Moreover, biomolecules potentially added to the scaffold can be partially removed during the leaching step in water.

1.3.2.3 Freeze Drying

An emulsion of polymer solution and water is prepared and rapidly cooled down to block the liquid structure in a solid one. Organic solvent and water are subsequently removed by freeze drying leaving a solid porous polymer with porosity up to 90% [96]. This technique has been applied to many biocompatible polymers such as PLA, PGA, PLAGA, PCL, chitosan and alginate [97–100]. Despite its versatility, freeze drying is a time and energy consuming method that takes several days to completely eliminate solvents.

1.3.2.4 Solid Freeform Fabrication

Solid freeform fabrication process is a computerized technique involving the design of a scaffold model through a CAD system which elaborates it as a series of cross sections [84, 101–103]. These sections are built by a rapid prototyping machine that lays down layers of material starting from the bottom and moving up a layer at a time to create the scaffold. This method enables to fabricate large and complex 3D objects with a precise control of pore architecture. Unfortunately, this top-down approach does not create nano-scaled structures and requires expensive equipments.

1.3.2.5 Thermally Induced Phase Separation

In this approach a polymer solution is cooled to low temperatures in order to induce a liquid–liquid or solid–liquid phase separation. The solvent located in the solvent-rich phase is subsequently removed by sublimation, leaving a porous polymer scaffold [104–106]. Scaffold morphology is controlled by the type of solvent, polymer concentration and phase separation temperature. This technique can be used to fabricate scaffolds from many types of polymers and polymeric composite materials, obtaining highly interconnected pores with tuneable pore size and also combined macro and microporous structures [107, 108]. The method enables to produce also nanofibrous networks with porosity of 98% and fibre diameters in the range 50–500 nm [3].

1.3.2.6 Peptide Self-Assembly

Peptides can be ad hoc synthesized to be able, in certain conditions, to generate $\alpha\beta$-helixes or β-sheets that self-assemble into stable nanofibres [109–111]. The mechanism is driven by non-covalent bonds and ionic interaction and it is controlled by pH and peptide concentration. These structures are used as synthetic ECM, even if, due to lack of mechanical strength, they are often used upon incorporation into more mechanically resilient scaffolds [112].

Two scaffold fabrication technologies, gas foaming and electrospinning, which were employed in the course of the present Ph.D., are illustrated in more detail in the following paragraphs.

1.3.2.7 Gas Foaming

Gas foaming has been commonly employed to produce microcellular foams of
thermoplastic polymers [113–116] such as polymethylmethacrylate and polystyrene, but only in 1994 Mooney et al. applied this method for the production of
$PLA_{50}GA_{50}$ scaffolds for TE [117]. Since then gas foaming has become an
appealing technique for fabricating microporous scaffolds [118, 119].

Supercritical carbon dioxide ($scCO_2$) is the most common substance employed,
which, once turned into gas phase, acts as a porogen to generate pores within a
polymer matrix. The method exploits the unique properties of $scCO_2$ that, combining liquid-like densities (high solvent power) with gas-like viscosities (high
diffusion rates) [118], is used for a wide range of applications in polymer synthesis, extraction and impregnation processes, particle formation and blending
[120, 121]. Moreover, CO_2 has a relatively low critical point ($T_c = 31$ °C,
$P_c = 7.4$ MPa) that can be easily achieved in a high-pressure equipped laboratory.

Basically, the method consists in dissolving $scCO_2$ in a solid polymer at high
pressure, generating a low viscosity mixture. Subsequently, the depressurization
decreases $scCO_2$ solubility in the polymer and leads to the phase transition of CO_2
from supercritical to gas. CO_2 bubble nucleation occurs and nuclei growth generates
pores within the polymer. Concomitantly, viscosity of the polymer-$scCO_2$ mixture
increases till all the gas has escaped from the polymer, leaving behind a solid structure
with "locked in" pores [122] (Fig. 1.6). Scaffold morphology can be controlled by
varying the amount of $scCO_2$ incorporated and its release rate from the polymer.

The main benefit of using $scCO_2$ foaming for the production of scaffolds is the
reduction of problems associated with residual solvents that can be toxic to
mammalian cells. The capability of producing scaffolds without the use of any
toxic solvent makes this technique unique and extremely interesting with respect
to all other scaffold fabrication methods known to date. However, it is well documented that foamed scaffolds can exhibit inadequate pore interconnectivity
especially at the scaffold surface, where a non-porous layer forms probably
because of the rapid diffusion of the CO_2 from the surface as the pressure is
released [123]. This issue has been overcome by combining gas foaming with the

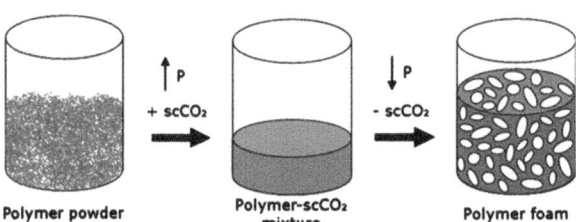

Fig. 1.6 Schematic description of $scCO_2$ foaming process. Polymer is firstly saturated with
$scCO_2$ at high pressure, generating a low viscous polymer-$scCO_2$ mixture. Then, the pressure
release produces a polymer foam

salt leaching procedure [117, 124] or by trimming off the non-porous skin from the scaffold after its fabrication [125, 126].

ScCO$_2$ foaming of biodegradable polymers has been mainly applied to poly-α-hydroxacids, though some publications also report the use of PCL [127, 128]. Given their morphology and their mechanical performances, these scaffolds can be employed for TE of hard tissues, such as bone and cartilage. For such applications PLA e PLAGA copolymers with different compositions have been successfully foamed either as plain materials [117, 125, 129] or in combination with inorganic substances [87, 130]. Moreover, scCO$_2$ foaming is a powerful technology for the production of scaffolds loaded with biomolecules such as growth factors [131–135] and DNA [134, 135] that, once released, have been shown to maintain their activity.

1.3.2.8 Electrospinning

Electrospinning (ES) enables the production of non-woven mats composed of sub-micrometric fibres from a polymer solution or melt. The process was invented at the beginning of 1900 by Cooley and Morton [136, 137] and it was mainly developed thanks to the contribution of Formhals' patents [138–142] for textile applications. However, only at the beginning of this century ES began to be used for scaffold fabrication [143–147] and nowadays it is considered a powerful technique that is employed to this aim by many researchers all around the world.

ES (Fig. 1.7) involves the application of a high voltage difference between a positively charged metallic needle, ejecting the polymeric solution, and a grounded collector. When the electrostatic force overcomes the cohesive force of the solution an electrically charged jet emerges from the capillary. The fluid jet is accelerated and stretched by the external electric field and thins dramatically while travelling towards the collector, leading to the formation of continuous solid fibres as the solvent evaporates. ES products appear as highly porous non-woven sheets made of submicrometric fibres with large surface area to volume ratio. Fibre diameter, morphology and arrangement in space can be controlled by a proper selection of the spinning parameters.

ES has been demonstrated to be an extremely versatile technology able to produce fibres with diameters ranging from few nanometers to several microns, from over two hundred synthetic and natural polymers, in the form of plain fibres, blends and organic–inorganic composite fibres [148–150]. Another advantage is the simplicity of the process that does not require any sophisticated and expensive equipment and that can be easily scaled-up for mass production.

Besides the aforementioned benefits of the process, its power arises from the morphological features of the products obtained. Indeed, ES fibre dimensions and spatial organization resemble the fibrous component of ECM, making ES a technology for the production of morphologically biomimetic scaffolds. As a consequence, this kind of scaffolds can be used to elicit different responses from the same cell phenotype only thanks to their particular topography. Indeed, it is well-established that, besides being influenced by external chemical signals

Fig. 1.7 Scheme of the ES process

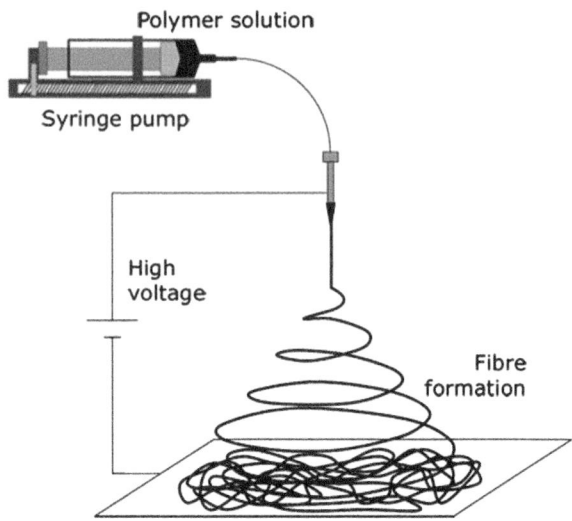

coming both from ECM and from nearby cells, cell behaviour is manipulated also by the morphological features of their environment that control cell adhesion, orientation, motility, gene expression, etc. [151]. A comprehensive review about the effect of surface topography on cells is provided by Stevens et al. [152]. The authors describe cell behaviour interacting with differently structured scaffolds and conclude that nanoscaled architectures promote better spreading and attachment when compared with microscaled scaffolds (Fig. 1.8). Their model was supported by several studies reporting that cells are able to better adhere and spread when cultured on sub-micrometric fibres with respect to micrometric ones [153, 154].

Given the many advantages in using ES technology, this is certainly one of the most extensively used techniques for scaffold fabrication. The scientific literature reports many successful applications of ES products for supporting different types of cells, demonstrating that these kinds of scaffolds are promising for the regeneration of several types of tissues (Table 1.2). Given the intrinsic mechanical properties of ES non-woven scaffolds, they are mainly applied for the reconstruction of soft tissues. As an example, the non-woven nanofibrous sheets obtained by ES are suitable for wound dressing and wound healing applications. The process also allows to obtain highly aligned fibres that are able to support and direct cells which are oriented in their natural environment, such as neuronal cells [155]. Moreover, thanks to the versatility of the technology, tubular scaffolds, which are ideally suited for replacing damaged blood vessels, can be easily fabricated [156]. ES scaffolds are also studied for hard TE (e.g. bone) even if, for these applications, polymeric materials are usually combined with inorganic substances in order to improve the biomimetic and mechanical features of the scaffold [157].

The high surface area to volume ratio of ES mats makes them also suitable for drug delivery applications. Many substances have been incorporated within fibres

Fig. 1.8 Cells binding to microstructured scaffolds flatten, exhibiting a morphology similar to that assumed on flat surfaces (such as 2D TCPS). Instead, nanostructured scaffolds offers many binding sites to cell membrane receptors and allow cells to assume a spread morphology more similar to that they have in natural ECM (figure adapted from [152], Copyright (2005), with permission from AAAS publications)

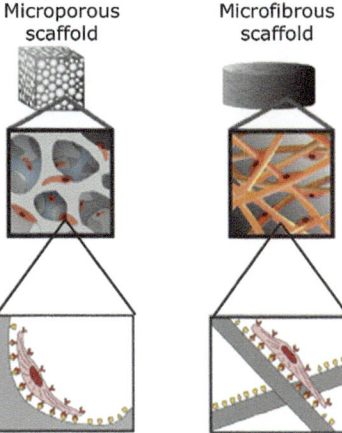 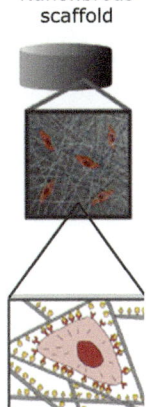

Table 1.2 Examples of ES biomaterials designed for TE applications

TE Application	Materials	References
Skin	PCL/collagen	[158, 159]
	Collagen	[156]
	Chitin	[154]
	Polystyrene	[160]
	Polyurethane	[161]
Nerve	PCL	[162, 163]
	PCL/collagen	[164]
	PLA	[165, 166]
	$PLA_{10}GA_{90}$	[167]
Bone	PCL	[168]
	PCL/β-TCP	[169]
	$PCL/CaCO_3$	[170]
	PLA/HA	[171]
	Gelatin/HA	[172]
Vascular graft	PCL/collagen/elastin	[173]
	PLA/collagen/elastin	[173, 174]
	$PLA_{80}CL_{20}$/collagen/elastin	[173]
	$PLA_{50}GA_{50}$/collagen/elastin	[173, 175]
	$PLA_{75}CL_{25}$	[59]
	$PLA_{50}CL_{50}$	[176]
	Collagen/eleastin	[177]
	PLA/PCL	[178]
	$PLA_{75}GA_{25}$/collagen	[179]
	Silk	[180]
Heart	PCL	[181]
	Pulyurethane	[182]
	$PLA/PLA_{75}GA_{25}/PLA_{10}GA_{90}$	[183]
Cartilage	PCL	[184]
	PCL/PLA	[185]

simply by adding them to polymer solution or through coaxial ES [186, 187]. The latter process consists in electrospinning two different solutions flowing through two capillaries, with the smaller one positioned inside the larger one. This configuration can encapsulate a smaller fibre within a larger one, leading to core–shell morphology. Drugs [188–192], proteins [193–197], growth factors [198, 199] and DNA [200] have been successfully incorporated inside polymeric ES fibres. It is worth noting that the capability to incorporate biologically relevant substances within scaffolds in a one-step process gives the possibility to combine, in a unique product, the function of supporting cell growth with the beneficial effects that arise from the interaction of cells with the delivered biomolecules.

1.4 Scaffold Functionalization

The need to address cell activity towards tissue regeneration has driven the scientific community to make efforts for designing bio-functionalized scaffolds. The main objective is to make the scaffold as much biomimetic as possible, not only in terms of 3D structural features, but also from a chemical point of view. To this aim the ECM is inevitably taken as a benchmark.

As previously discussed in Sect. 1.2, ECM and the hosted cells communicate through specific proteins either located at the surface of the ECM fibrous component (e.g. laminin, fibronectin, vitronectin, etc.) or dissolved in the gel-like ECM component (e.g. growth factors). The incorporation of these proteins (or of peptides displaying similar functionalities) is considered a valid approach to confer bioactivity to the scaffold and it is commonly achieved through either bulk or surface functionalization. In the bulk functionalization, biomolecules are incorporated within the polymer matrix and they are released in the surrounding environment either by diffusion or via scaffold degradation. Growth factors (GFs) are probably the most widely used biomolecules in such applications [201]; the role of these proteins is to transmit signals controlling cell migration, differentiation, and proliferation. A list of the most commonly employed GFs together with their known activities is provided by Chen and Mooney [202]. Other molecules that have been incorporated within scaffold matrixes are antibiotics, anticancer drugs, anti-inflammatory dugs and DNA plasmids [203].

Surface modification is the other approach aimed at conferring bioactivity to scaffolds since biomaterial surface properties regulate the interaction with the physiological environment and with cells. In particular, the scientific community has dedicated strong efforts in modifying biomaterial to stimulate cell adhesion by binding specific peptides sequences to biomaterial surface. RGD sequence (R: arginine; G: glycine; D: aspartic acid) has been largely employed for stimulating cell attachment on synthetic surfaces since it was identified as the minimal sequence required to mediate cell adhesion in most ECM proteins. RGD can be introduced in the biomaterial either by synthesizing polymers containing RGD sequences in the macromolecular chains [204] or via post-fabrication treatments for attaching RGD sequences to material surfaces. In this context, films have been mostly employed whereas modification of scaffold surface with these biomolecules is less explored [204].

In the course of the present Ph.D. both fuctionalization approaches were attempted. Bulk functionalization was achieved through the incorporation of a GF in ES fibres whereas surface functionalization of ES fibres was carried out by Surface Initiated Atom Transfer Radical Polymerization (SI-ATRP).

References

1. Mason C, Dunnill P (2008) A brief definition of regenerative medicine. Regen Med 3:1
2. Vacanti JP, Vacanti CA (2007) The history and scope of tissue engineering. In: Lanza R, Langer R, Vacanti J (eds) Principles of Tissue Engineering. Elsevier, San Diego, pp 3–6
3. Langer R, Vacanti JP (1993) Tissue engineering. Science 260:920
4. Leor J, Amsalem Y, Cohen S (2005) Cells, scaffolds and molecules for myocardial tissue engineering. Pharmacol Ther 105:151
5. Stock UA, Vacanti JP (2001) Tissue engineering: current state and prospects. Ann Rev Med 52:443
6. Place ES, Evans ND, Stevens MM (2009) Complexity in biomaterials for tissue engineering. Nat Mater 8:457
7. Slack J (2007) Molecular biology of the cell. In: Lanza R, Langer R, Vacanti J (eds) Principles of Tissue Engineering. Elsevier, San Diego, pp 53–66
8. Bilodeau K, Mantovani D (2006) Bioreactors for tissue engineering: focus on mechanical constrains. A comparative review. Tissue Eng 12:2367
9. Vunjak-Novakovic G, Freed LE, Biron RJ, Langer R (1996) Effects of mixing on the composition and morphology of tissue-engineered cartilage. Bioengineering Food Nat Prod 42:850
10. Martin I, Wendt D, Heberer M (2004) The role of bioreactors in tissue engineering. Trends Biotechnol 22:80
11. Folkman J, Moscona A (1978) Role of cell shape in growth control. Nature 273:345
12. Lutolf MP, Hubbell JA (2005) Synthetic biomaterials as instructive extracellular microenvironments for morphogenesis in tissue engineering. Nat Biotechnol 23:47
13. Giancotti FG (2000) Complexity and specificity of integrin signaling. Nat Cell Biol 2:E13
14. Bissell MJ, Hall HG, Parry G (1982) How does the extracellular matrix direct gene expression? J Theor Biol 99:31
15. Streuli CH, Bailey N, Bissell MJ (1991) Control of mammary epithelial differentiation: basement membrane induces tissue-specific gene expression in the absence of cell-cell interaction and morphological polarity. J Cell Biol 115:1383
16. Chen CS, Mrksich M, Huang S, Whitesides GM, Ingber DE (1997) Geometric control of cell live and death. Science 276:1425
17. Blaschke RJ, Howlett AR, Desprez P-Y, Peterson OW, Bissell MJ (1994) Cell differentiation by extracellular matrix components. Methods Enzymol 245:535
18. Veiseh M, Turley EA, Bissell MJ (2008) Top–down analysis of a dynamic environment: extracellular matrix structure and function. In: Laurencin CT, Nair LS (eds) Nanotechnology and Tissue Engineering: The Scaffold. CRC Press/Taylor & Francis Group, Boca Raton, pp 33–51
19. ASTM F2150-07 (2007) Standard guide for characterization and testing of biomaterial scaffolds used in tissue-engineered medical products
20. Salgado AJ, Coutinho OP, Reis RL (2004) Bone tissue engineering: state of the art and future trends. Macromol Biosci 4:743

21. Ma PX (2004) Scaffolds for tissue fabrication. Materials Today 7:30
22. Yoon DM, Fisher JP (2007) Polymeric scaffolds for tissue engineering applications. In: Fisher JP, Mikos AG, Bronzino JD (eds) Tissue Engineering. CRC Press/Taylor & Francis Group, Boca Raton, pp 8-1–8-18
23. Karande TS, Agrawal CM (2008) Functions and requirments of synthetic scaffolds in tissue engineering. In: Laurencin CT, Nair LS (eds) Nanotechnology and Tissue Engineering: The Scaffold. CRC Press/Taylor & Francis Group, Boca Raton, pp 53–86
24. MUschler GF, Nakamoto C, Griffith LG (2004) Engineering principles of clinical cell-based tissue engineering. J Bone Joint Surg 86-A:1541
25. Vert M, Li SM, Spenlehauer G, Guerin P (1992) Bioresorbability and biocompatibility of aliphatic polyesters. J Mater Sci Mater Med 3:432
26. Hutmacher DW (2000) Scaffolds in tissue engineering bone and cartilage. Biomaterials 21:2529
27. Muschler GF, Nakamoto C, Griffith LG (2004) Engineering principles of clinical cell-based tissue engineering. J Bone Joint Surg 86A:1541
28. Van der Flier A, Sonnenberg A (2001) Function interactions of integrins. Cell Tissue Res 305:285
29. Plow EF, Haas TA, Zhang L, Loftus J, Smith JW (2000) Ligand binding to integrins. J Biol Chem 275:21785
30. Roach P, Farrar D, Perry CC (2005) Interpretation of protein adsorption: surface-induced conformational changes. J Am Chem Soc 127:8168
31. Mathieu HJ (2001) Bioengineered material sufaces for medical applications. Surf Interface Anal 32:3
32. Saltzman WM, Kyriakides TR (2007) Cell interactions with polymers. In: Lanza R, Langer R, Vacanti J (eds) Principles of Tissue Engineering. Elsevier, San Diego, pp 279–296
33. Jiao Y-P, Cui F-Z (2007) Surface modification of polyester biomaterials for tissue engineering. Biomed Mater 2:R24
34. Karande TS, Ong JL, Agrawal CM (2004) Diffusion in musculoskeletal tissue engineering scaffolds: design issues related to porosity, permeability, architecture, and nutrient mixing. Ann Biomed Eng 32:1728
35. Kim B-S, Mooney DJ (1998) Development of biocompatible synthetic extracellular matrices for tissue engineering. Trends Biotechnol 16:224
36. Williams DF (1987) Definitions in biomaterials. In: Proceeding of a Consensus Conference of the European Society for Biomaterials. Elsevier, Amsterdam
37. Park JB (2000) Biomaterials. In: Bronzino JD (ed) The Biomedical Engineering Handbook. CRC Press LLC, Boca Raton, p IV-1
38. Hench LL, Polak JM (2002) Third-generation biomedical materials. Science 295:1014
39. Kim S-S, Park MS, Jeon O, Choi CY, Kim B-S (2006) Poly(lactide-co-glycolide)/hydroxyapatite composite scaffolds for bone tissue engineering. Biomaterials 27:1399
40. Zhang R, Ma PX (1999) Poly(α-hydroxyl acids)/hydroxyapatite porous composites for bone-tissue engineering. I. Preparation and morphology. J Biomed Mater Res 44:446
41. Rezwan K, Chen QZ, Blacker JJ, Boccaccini AR (2006) Biodegradable and bioactive porous polymer/inorganic composite scaffolds for bone tissue engineering. Biomaterials 27:3413
42. Ikada Y (2006) Tissue engineering: fundamentals and applications. Elsevier, Amsterdam
43. Caterson EJ, Nesti LJ, Li W-J, Danielson KG, Albert TJ, Vaccaro AR, Tuan RS (2001) Three-dimensional cartilage formation by bone marrow-derived cells seeded in polylactide/alginate amalgam. J Biomed Mater Res 57:394
44. Zhang Y, Zhang M (2001) Synthesis and characterization of macroporous chitosan/calcium phosphate composite scaffolds for tissue engineering. J Biomed Mater Res 55:304
45. Kim H-W, Kim H-E, Salih V (2005) Stimulation of osteoblast responses to biomimetic nanocomposites of gelatin-hydroxyapatite for tissue engineering scaffolds. Biomaterials 26:5221

46. Chen G, Sato T, Ushida T, Hirochika R, Shirasaki Y, Ochiai N, Tateishi T (2003) The use of a novel PLGA fiber/collagen composite web as a scaffold for engineering of articular cartilage tissue with adjustable thikness. J Biomed Mater Res 67:1170

47. Sarasam AR, Samli AI, Hess L, IHnat MA, Madihally SV (2007) Blending chitosan with polycaprolactone: porous scaffolds and toxicity. Macromol Biosci 7:1160

48. Agrawal CM, Ray RB (2001) Biodegradable polymeric scaffolds for musculoskeletal tissue engineering. J Biomed Mater Res 55:141

49. Seal BL, Otero TC, Panitch A (2001) Polymeric biomaterials for tissue and organ regeneration. Mater Sci Eng R 34:147

50. Agrawal CM, Athanasiou KA (1997) Technique to control pH in vicinity of biodegrading PLA-PGA implants. J Biomed Mater Res 38:105

51. Morita S-I, Ikada Y (2002) Lactide copolymers for scaffolds in tissue engineering. In: Tissue engineering and biodegradable equivalents: scientific and clinical applications, pp 111–122

52. Athanasiou KA, Agrawal CM, Barber FA, Burkhart SS (1998) Orthopaedic applications for PLA-PGA biodegradable polymers. Arthrosc J Arthrosc Relat Surg 14:726

53. Gilding DK, Reed AM (1979) Biodegradable polymers for use in surgery–polyglycolic/poly(actic acid) homo- and copolymers. Polymer 20:1459

54. Li S (1999) Hydrolytic degradation characteristics of aliphatic polyesters derived from lactic and glycolic acid. J Biomed Mater Res 48:342

55. Vert M, Garreau H, Maudit J, Boustta M, Schwach G, Engel R, Coudane J (1997) Complexity of the hydrolitic degradation of aliphatic polyesters. Die Angew Makrom Chem 247:239

56. Li S, Garreau H, Vert M (1990) Structure-property relationships in the case of the degradation of massive aliphatic poly-(α-hydroxy acids) in aqueous media, part 2:degradation of lactide-glycolide copolymers: PLA37.5GA25 and PLA75GA25. J Mater Sci Mater Med 1:131

57. Miller RA, Brady JM, Cutright DE (1977) Degradation rate of oral resorbable implants (polylactate and polyglycolate): rate modification with changes in PLA/PGA copolymer ratios. J Biomed Mater Res 11:711

58. De Groot JH, Zijlstra FM, Kuipers HW, Pennings AJ, Klompmaker J, Veth RPH, Jansen HWB (1997) Meniscal tissue regeneration in porous 50/50 copoly(L-lactide/ε-caprolactone) implants. Biomaterials 18:613

59. Xu CY, Inai R, Kotaki M, Ramakrishna S (2004) Aligned biodegradable nanofibrous structure: a potential scaffold for blood vessel engineering. Biomaterials 25:877

60. Mo XM, Xu CY, Kotaki M, Ramakrishna S (2004) Electrospun P(LLA-CL) nanofiber: a biomimetic extracellular matrix for smooth muscle cell and endothelial cell proliferation. Biomaterials 25:1883

61. Welle A, Kroger M, Doring M, Niederer K, Pindel E, Chronakis IS (2007) Electrospun aliphatic polycarbonates as tailored tissue scaffold materials. Biomaterials 28:2211

62. Mukherjee DP, Smith DF, Rogers SH, Emmanual JE, Jadin KD, Hayes BK (2009) Effect of 3D-microstructure of bioabsorbable PGA:TMC scaffolds on the growth of chondrogenic cell. J Biomed Mater Res Part B Appl Biomater 88B:92

63. Vinoy T, Zhang X, Catledge SA, Vohra YK (2007) Functionally graded electrospun scaffolds with tunable mechanical properties for vascular tissue regeneration. Biomed Mater 2:224

64. Pego AP, Siebum B, Luyn V, Gallego XJ, Seijen YV, Poot AA, Grijpma DW, Feijen J (2003) Preparation of degradable porous structures based on 1, 3-trimethylene carbonate and D, L-lactide (co)polymers for heart tissue engineering. Tissue Eng 9:981

65. Pego AP, Poot AA, Grijpma DW, Feijen J (2003) Biodegradable elastomeric scaffolds for soft tissue engineering. J Controlled Release 87:69

66. Plikk P, Malberg S, Albertsson A-C (2009) Design of resorbable porous tubular copolyester scaffolds for use in nerve regeneration. Biomacromolecules 10:1259

67. Martina M, Hutmacher DW (2007) Biodegradable polymers applied in tissue engineering research: a review. Polym Int 56:145
68. Ambrosio AM, Allcock HR, Katti DS, Laurencin CT (2002) Degradable polyphosphazene/poly(hydroxyester) blends: degradation studies. Biomaterials 23:1667
69. Allcock HR, Fuller TJ, Matsumura K (1982) Hydrolysis pathways for amminophosphazenes. Inorg Chem 21:515
70. Laurencin CT, El-Amin SF, Ibim SE, Willoughby DA, Attawia M, Allcock HR, Ambrosio AM (1996) A highly porous 3-dimensional polyphosphazene polymer matrix for skeletal tissue regeneration. J Biomed Mater Res 30:133
71. Conconi MT, Lora S, Baiguera S, Boscolo E, Folin M, Scienza R, Rebuffat P, Parnigotto PP, Nussdorfer GG (2004) In vitro culture of rat neuromicrovascular endothelial cells on polymeric scaffolds. J Biomed Mater Res 71A:669
72. Ambrosio AM, Sahota JS, Runge C, Kurtz SM, Lakshmi S, Allcock HR, Laurencin CT (2003) Novel polyphosphazene-hydroxyapatite composites as biomaterials. IEEE Eng Med Biol Mag 22:18
73. Temenoff JS, Mikos AG (2000) Injectable biodegradable materials for orthopedic tissue engineering. Biomaterials 21:2405
74. Payne RG, McGonigle JS, Yaszemski Mj, Yasko AW, Mikos AG (2002) Development of an injectable, in situ crosslinkable, degradable polymeric carrier for osteogenic cell populations Part 3 Proliferation, differentiation of encapsulated marrow stromal osteoblasts cultured on crosslinking poly(propylene fumarate). Biomaterials 23:4381
75. Yaszemski Mj, Payne RG, Hayes WC, Langer R, Aufdemorte TB, Mikos AG (1995) The ingrowth of new bone tissue, initial mechanical properties of a degrading polymeric composite scaffold. Tissue Eng 1:41
76. Sundback CA, Shyu JY, Wang Y, Faquin WC, Langer R, Vacanti J, Hadlock TA (2005) Biocompatibility analysis of poly(glycerol sebacate) as a nerve guide material. Biomaterials 26:5454
77. Gao J, Ensley AE, Nerem RM, Wang Y (2007) Poly(glycerol sebacate) supports the proliferation and phenotypic protein expression of primary baboon vascular cells. J Biomed Mater Res 83A:1070
78. Gao J, Crapo PM, Wang Y (2006) Macroporous elastomeric scaffolds with extensive micropores for soft tissue engineering. Tissue Eng 12:917
79. Wang Y, Ameer GA, Sheppard BJ, Langer R (2002) A tough biodegradable elastomer. Nat Biotechnol 20:602
80. Smith LA, Ma PX (2004) Nano-fibrous scaffolds for tissue engineering. Colloids Surf B Biointerfaces 39:125
81. Murphy MB, Mikos AG (2007) Polymer scaffold fabrication. In: Lanza R, Langer R, Vacanti J (eds) Principles of Tissue Engineering. Elsevier, San Diego, pp 309–321
82. Katoh K, Tanabe T, Yamauchi K (2004) Novel approach to fabricate keratin sponge scaffolds with controlled pore size and porosity. Biomaterials 25:4255
83. Kang HW, Tabata Y, Ikada Y (1999) Fabrication of porous gelatin scaffolds for tissue engineering. Biomaterials 20:1339
84. Leong KF, Cheah CM, Chua CK (2003) Solid freeform fabrication of three-dimensional scaffolds for engineering replacement tissues and organs. Biomaterials 24:2363
85. Ma PX, Zhang R (1999) Synthetic nano-scale fibrous extracellular matrix. J Biomed Mater Res 46:60
86. Woolfson DN, Ryadnov MG (2006) Peptide-based fibrous biomaterials: some things old, new and borrowed. Curr Opin Chem Biol 10:559
87. Mathieu L, Montjovent M-O, Bourban P-E, Pioletti DP, Manson J-AE (2005) Bioresorbable composites prepared by supercritical fluid foaming. J Biomed Mater Res Part A 75:89
88. Ma K, Chan CK, Liao S, Hwang WYK, Feng Q, Ramakrishna S (2008) Electrospun nanofiber scaffolds for rapid and rich capture of bone marrow-derived hematopoietic stem cells. Biomaterials 29:2096

89. Grande DA, Halberstadt C, Naughton G, Schwartz R, Manji R (1997) Evaluation of matrix scaffolds for tissue engineering of articular cartilage grafts. J Biomed Mater Res 34:211

90. Sittinger M, Reitzel D, Dauner M, Hierlemann H, Hammer C, Kastenbauer E, Planck H, Burmester GR, Bujia J (1996) Resorbable polyesters in cartilage engineering: affinity and biocompatibility of polymer fiber structures to chondrocytes. J Biomed Mater Res Part B Appl Biomater 33:57

91. Ma PX, Langer R (1999) Morphology and mechanical function of long-term in vitro engineered cartilage. J Biomed Mater Res 44:217

92. Mikos AG, Thorsen AJ, Czerwonka LA, Bao Y, Langer R, Winslow DN, Vacanti JP (1994) Preparation and characterization of poly(-lactic acid) foams. Polymer 35:1068

93. Holy CE, Dang SM, Davies JE, Shoichet MS (1999) In vitro degradation of a novel poly(lactide-co-glycolide) 75/25 foam. Biomaterials 20:1177

94. Suh SW, Shin YJ, Kim J, Min CH, Beak CH, Kim D-I, Kim H, Jeon SS, Choo IW (2002) Effect of different particles on cell proliferation in polymer scaffolds using a solvent-casting and particulate leaching technique. ASAIO J 48:460

95. Wake MC, Gupta PK, Mikos AG (1996) Fabrication of pliable biodegradable polymer foams to engineer soft tissues. Cell Transplant 5:465

96. Whang K, Thomas H, Healy KE, Nuber G (1995) A novel method to fabricate bioabsorbable scaffolds. Polymer 36:837

97. Whang K, Goldstick TK, Healy KE (2000) A biodegradable polymer scaffold for delivery of osteotropic factors. Biomaterials 21:2545

98. Shen F, Cui YL, Yang LF, Yao KD, Dong XH, Jia WY, Shi HD (2000) A study on the fabrication of porous chitosan/gelatin network scaffold for tissue engineering. Polym Int 49:1596

99. Hou Q, Grijpma DW, Feijen J (2003) Preparation of interconnected highly porous polymeric structures by a replication and freeze-drying process. J Biomed Mater Res Part B Appl Biomater 67B:732

100. Ho M-H, Kuo P-Y, Hsieh H-J, Hsien T-Y, Hou L-T, Lai J-Y, Wang D-M (2004) Preparation of porous scaffolds by using freeze-extraction and freeze-gelation methods. Biomaterials 25:129

101. Hollister SJ (2005) Porous scaffold design for tissue engineering. Nat Mater 4:518

102. Yan Y, Xiong Z, Hu Y, Wang S, Zhang R, Zhang C (2003) Layered manufacturing of tissue engineering scaffolds via multi-nozzle deposition. Mater Lett 57:2623

103. Hutmacher DW, Sittinger M, Risbud MV (2004) Scaffold-based tissue engineering: rationale for computer-aided design and solid free-form fabrication systems. Trends Biotechnol 22:354

104. Lo H, Kadiyala S, Guggino SE, Leong KW (1996) Poly(L-lactic acid) foams with cell seeding and controlled-release capacity. J Biomed Mater Res 30:475

105. Blaker JJ, Maquet V, Jerome R, Boccaccini AR, Nazhat SN (2005) Mechanical properties of highly porous PDLLA/Bioglass composite foams as scaffolds for bone tissue engineering. Acta Biomater 1:643

106. Schugens C, Maquet V, Grandfils C, Jerome R, Teyssie P (1996) Polylactide macroporous biodegradable implants for cell transplantation. II. Preparation of polylactide foams by liquid-liquid phase separation. J Biomed Mater Res 30:449

107. Nam YS, Park TG (1999) Porous biodegradable polymeric scaffolds prepared by thermally induced phase separation. J Biomed Mater Res 47:8

108. Nam YS, Park TG (1999) Biodegradable polymeric microcellular foams by modified thermally induced phase separation method. Biomaterials 20:1783

109. Holmes TC (2002) Novel peptide-based biomaterial scaffolds for tissue engineering. Trends Biotechnol 20:16

110. Beniash E, Hartgerink JD, Storrie H, Stendhal JC, Stupp SI (2005) Self-assembling peptide amphiphile nanofiber matrices for cell entrapment. Acta Biomater 1:387

111. Yokoi H, Kinoshita T, Zhang S (2005) Dynamic reassembly of peptide RADA16 nanofiber scaffold. Proc Nat Acad Sci U S A 102:8414

112. Harrington DA, Cheng EY, Guler MO, Lee LK, Donovan JL, Claussen RC, Stupp SI (2006) Branched peptide-amphiphiles as self-assembling coatings for tissue engineering scaffolds. J Biomed Mater Res 78A:157

113. Goel SK, Beckman EJ (1994) Generation of microcellular polymeric foams using supercritical carbon dioxide. I: effect of pressure and temperature on nucleation. Polym Eng Sci 34:1137

114. Arora KA, Lesser AJ, McCarthy TJ (1998) Preparation and characterization of microcellular polystyrene foams processed in supercritical carbon dioxide. Macromolecules 31:4614

115. Colton JS, Suh NP (1987) The nucleation of microcellular thermoplastic foam with additives part I: theoretical considerations. Polym Eng Sci 27:485

116. Kumar V, Suh NP (1990) A process for making microcellular thermoplastic parts. Polym Eng Sci 30:1323

117. Mooney DJ, Baldwin DF, Suh NP, Vacanti JP, Langer R (1996) Novel approach to fabricate porous sponges of poly(D, L-lactic-co-glycolic acid) without the use of organic solvents. Biomaterials 17:1417

118. Woods HM, Silva MCG, Nouvel C, Shakesheff KM, Howdle SM (2004) Materials processing in supercritical carbon dioxide: surfactants, polymers and biomaterials. J Mater Chem 14:1663

119. Quirk RA, France RM, Shakesheff KM, Howdle SM (2004) Supercritical fluid technologies and tissue engineering scaffolds. Curr Opinion Solid State Mater Sci 8:313

120. Tomasko DL, Li H, Liu D, Han X, Wingert MJ, Lee LJ, Koelling KW (2003) A review of CO_2 applications in the processing of polymers. Ind Eng Chem Res 42:6431

121. Kazarian SG (2000) Polymer processing with supercritical fluids. Polym Sci Ser C 42:78

122. Barry JJA, Silva MMCG, Popov VK, Shakesheff KM, Howdle SM (2006) Supercritical carbon dioxide: putting the fizz into biomaterials. Philos Trans Roy Soc A 364:249

123. Goel SK, Beckman EJ (1994) Generation of microcellular polymeric foams using supercritical carbon dioxide. II: cell growth and skin formation. Polym Eng Sci 34:1148

124. Murphy WL, Dennis RG, Kileny JL, Mooney DJ (2002) Salt fusion: an approach to improve pore interconnectivity within tissue engineering scaffolds. Tissue Eng 8:43

125. Tai H, Mather ML, Howard D, Wang W, White LJ, Crowe JA, Morgan SP, Williams DJ, Howdle SM, Shakesheff KM (2007) Control of pore size and structure of tissue engineering scaffolds produced by supercritical fluid processing. Eur Cell Mater 14:64

126. Barry JJA, Gidda HS, Scotchford CA, Howdle SM (2004) Porous methacrylate scaffolds: supercritical fluid fabrication and in vitro chondrocyte responses. Biomaterials 25:3559

127. Jenkins MJ, Harrison KL, Silva MCG, Whitaker MJ, Shakesheff KM, Howdle SM (2006) Characterization of microcellular foams produced from semi-crystalline PCL using supercritical carbon dioxide. Eur Polym J 42:3145

128. Xu Q, Ren X, Chang Y, Wang J, Yu L, Dean K (2004) Generation of microcellular biodegradable polycaprolactone foams in supercritical carbon dioxide. J Appl Polym Sci 94:593

129. Singh L, Kumar V, Ratner BD (2004) Generation of porous microcellular 85/15 poly (dl-lactide-co-glycolide) foams for biomedical applications. Biomaterials 25:2611

130. Mathieu LM, Mueller TL, Bourban P-E, Pioletti DP, Muller R, Manson J-AE (2006) Architecture and properties of anisotropic polymer composite scaffolds for bone tissue engineering. Biomaterials 27:916

131. Sheridan MH, Shea LD, Peters MC, Mooney DJ (2000) Bioabsorbable polymer scaffolds for tissue engineering capable of sustained growth factor delivery. J Controlled Release 64:91

132. Yang XB, Whitaker MJ, Sebald W, Clarke N, Howdle SM, Shakesheff KM, Oreffo ROC (2004) Human osteoprogenitor bone formation using encapsulated bone morphogenetic protein 2 in porous polymer scaffolds. Tissue Eng 10:1037

133. Hile DD, Amipour L, Akgerman A, Pishko MV (2000) Active growth factor delivery from poly(D, L-lactide-co-glycolide) foams prepared in supercritical CO2. J Controlled Release 66:177

134. Heyde M, Partridge KA, Howdle SM, Oreffo ROC, Garnett MC, Shakesheff KM (2007) Development of a slow non-viral DNA release system from PDLLA scaffolds fabricated using a supercritical CO_2 technique. Biotechnol Bioeng 98:679
135. Jang J-H, Shea LD (2003) Controllable delivery of non-viral DNA from porous scaffolds. J Controlled Release 86:157
136. Cooley JF (1902) Apparatus for electrically dispersing fluids. US patent 692631
137. Morton WJ (1902) Method of dispersing fluids. US patent 705691
138. Formhals A (1934) Process and apparatus for preparing artificial threads. US patent 1975504
139. Formhals A (1939) Method and apparatus for spinning. US patent 2160962
140. Formhals A (1940) Artificial thread and method of producing same. US patent 2187306
141. Formhals A (1943) Production of artificial fibers from fiber forming liquids. US patent 2323025
142. Formhals A (1944) Method and apparatus for spinning. US patent 2349950
143. Boland ED, Wnek GE, Simpson DG, Pawlowski KJ, Bowlin G (2001) Tailoring tissue engineering scaffolds using electrostatic processing techniques: a study of poly(glygolic acid) electrospinning. J Macromol Sci Pure Appl Chem A38:1231
144. Matthews JA, Wnek GE, Simpson DG, Bowlin GL (2002) Electrospinning of collagen nanofibers. Biomacromolecules 3:232
145. Li WJ, Laurencin CT, Caterson EJ, Tuan RS, Ko FK (2002) Electrospun nanofibrous structure: a novel scaffold for tissue engineering. J Biomed Mater Res 60:613
146. Boland ED, Bowlin GL, Simpson DG, Wnek GE (2001) Electrospinning of tissue engineering scaffolds. Polym Mater Sci Eng 85:51
147. Huang L, Nagapudi K, Apkarian RP, Chaikof EL (2001) Engineered collagen-PEO nanofibers and fabrics. J Biomater Sci Polym Edition 12:979
148. Liang D, Hsiao BS, Chu B (2007) Functional electrospun nanofibrous scaffolds for biomedical applications. Adv Drug Deliv Rev 59:1392
149. Chew SY, Wen Y, Dzenis Y, Leong KW (2006) The role of electrospinning in the energing field of nanomedicine. Curr Pharm Des 12:4751
150. Shiffman JD, Schauer CL (2008) A review: electrospinning of biopolymer nanofibres and their applications. Polym Rev 48:317
151. Webster TJ, Schadler LS, Siegel RW, Bizios R (2001) Mechanisms of enhanced osteoblast adhesion on nanophase alumina involve vitronectin. Tissue Eng 7:291
152. Stevens MM, George JH (2005) Exploring and engineering the cell surface interface. Science 310:1135
153. Kwon IK, Kidoaki S, Matsuda T (2005) Electrospun nano- to microfiber fabrics made of biodegradable copolyesters: structural characteristics, mechanical properties and cell adhesion potential. Biomaterials 26:3929
154. Noh HK, Lee SW, Kim JM, Oh JE, Kim KH, Chung CP, Choi SC, Park WH, Min BM (2006) Electrospinning of chitin nanofibers: degradation behavior and cellular response to normal human keratinocytes and fibroblasts. Biomaterials 27:3934
155. Cao H, Liu T, Chew SY (2009) The application of nanofibrous scaffolds in neural tissue engineering. Adv Drug Deliv Rev 61:1055
156. Sell SA, McClure MJ, Garg K, Wolfe PS, Bowlin GL (2009) Electrospinning of collagen/biopolymers for regenerative medicine and cardiovascular tissue engineering. Adv Drug Deliv Rev 61:1007
157. Jang JH, Castano O, Kim HW (2009) Electrospun materials as potential platforms for bone tissue engineering. Adv Drug Deliv Rev 61:1065
158. Chong EJ, Phan TT, Lim IJ, Zhang YZ, Bay BH, Ramakrishna S, Lim CT (2007) Evaluation of electrospun PCL/gelatin nanofibrous scaffold for wound healing and layered dermal reconstruction. Acta Biomater 3:321
159. Venugopal JR, Zhang Y, Ramakrishna S (2006) In vitro culture of human dermal fibroblasts on electrospun polycaprolactone collagen nanofibrous membrane. Artif Organs 30:440

160. Sun T, Mai SM, Norton D, Haycock JW, Ryan AJ, MacNeil S (2005) Self-organization of skin cells in three-dimensional electrospun polystyrene scaffolds. Tissue Eng 11:1023
161. Khil M-S, Cha D-I, Kim H-Y, Kim I-S, Bhattarai N (2003) Electrospun nanofibrous polyurethane membrane as wound dressing. J Biomed Mater Res Part B Appl Biomater 67B:675
162. Chew SY, Mi R, Hoke A, Leong KW (2009) The effect of the alignment of electrospun fibrous scaffolds on Schwann cell maturation. Biomaterials 29:653
163. Chew SY, Mi R, Hoke A, Leong KW (2007) Aligned protein-polymer composite fibers enhance nerve regeneration: a potential tissue-engineering platform. Adv Funct Mater 17:1288
164. Schnell E, Klinkhammer K, Balzer S, Brook G, Klee D, Dalton P, Mey J (2007) Guidance of glial cell migration and axonal growth on electrospun nanofibers of poly-ε-caprolactone and a collagen/poly-ε-caprolactone blend. Biomaterials 28:3012
165. Yang F, Murugan R, Wang S, Ramakrishna S (2005) Electrospinning of nano/micro scale poly(L-lactic acid) aligned fibers and their potential in neural tissue engineering. Biomaterials 26:2603
166. Koh HS, Yong T, Chan CK, Ramakrishna S (2008) Enhancement of neurite outgrowth using nano-structured scaffolds coupled with laminin. Biomaterials 29:3574
167. Bini TB, Wang S, Gao S, Ramakrishna S (2006) Poly(l-lactide-co-glycolide) biodegradable microfibers and electrospun nanofibers for nerve tissue engineering: an in vitro study. J Mater Sci 41:6453
168. Yoshimoto H, Shin YM, Terai H, Vacanti JP (2003) A biodegradable nanofiber scaffold by electrospinning and its potential for bone tissue engineering. Biomaterials 24:2077
169. Erisken C, Kalyon DM, Wang H (2008) Functionally graded electrospun polycaprolactone and β-tricalcium phosphate nanocomposites for tissue engineering applications. Biomaterials 29:4065
170. Fujihara K, Kotaki M, Ramakrishna S (2005) Guided bone regeneration membrane made of polycaprolactone/calcium carbonate composite nano-fibers. Biomaterials 26:4139
171. Sui G, Yang X, Mei F, Hu X, Chen G, Deng X, Ryu S (2007) Poly-L-lactic acid/ hydroxyapatite hybrid membrane for bone tissue regeneration. J Biomed Mater Res 82A:445
172. Kim H-W, Song J-H, Kim H-E (2005) Nanofiber generation of gelatin-hydroxyapatite biomimetics for guided tissue regeneration. Adv Funct Mater 15:1988
173. Lee SJ, Yoo JJ, Lim GJ, Atala A, Stitzel J (2007) In vitro evaluation of electrospun nanofiber scaffolds for vascular graft application. J Biomed Mater Res 83:999
174. Stitzel J, Pawlowski KJ, Wnek GE, Simpson DG, Bowlin GL (2001) Arterial smooth muscle cell proliferation on a novel biomimicking, biodegradable vascular graft scaffold. J Biomater Appl 16:22
175. Stitzel J, Lee SJ, Komura M, Berry J, Soker S, Lim G, Dyke MV, Czerw R, Yoo JJ, Atala A (2006) Controlled fabrication of a biological vascular substitute. Biomaterials 27:1088
176. Inoguchi H, Kwon IK, Inoue E, Takamizawa K, Maehara Y, Matsuda T (2006) Mechanical responses of a compliant electrospun poly(L-lactide-co-ε-caprolactone) small-diameter vascular graft. Biomaterials 27:1470
177. Boland ED, Matthews JA, Pawlowski KJ, Simpson DG, Wnek GE, Bowlin GL (2004) Electrospinning collagen and elastin: preliminary vascular tissue engineering. Front Biosci 9:1422
178. Vaz CM, van Tuijl S, Bouten CVC, Baaijens FPT (2005) Design of scaffolds for blood vessel tissue engineering using a multi-layering electrospinning technique. Acta Biomater 1:575
179. In Jeong S, Kim SY, Cho SK, Chong MS, Kim KS, Kim H, Lee SB, Lee YM (2007) Tissue-engineered vascular grafts composed of marine collagen and PLGA fibers using pulsatile perfusion bioreactors. Biomaterials 28:1115

180. Soffer L, Wang X, Zhang X, Kluge J, Dorfmann L, Kaplan DL, Leisk G (2009) Silk-based electrospun tubular scaffolds for tissue-engineered vascular grafts. J Biomater Sci Polym Ed 19:653
181. Shin M, Ishii O, Sueda T, Vacanti JP (2004) Contractile cardiac grafts using a novel nanofibrous mesh. Biomaterials 25:3717
182. Rockwood DN, Akins RE, Parrag IC, Woodhouse KA, Rabolt JF (2008) Culture on electrospun polyurethane scaffolds decreases atrial natriuretic peptide expression by cardiomyocytes in vitro. Biomaterials 29:4783
183. Zong X, Bien H, Chung CY, Yin L, Fang D, Hsiao BS, Chu B, Entcheva E (2005) Electrospun fine-textured scaffolds for heart tissue constructs. Biomaterials 26:5330
184. Li WJ, Tuli R, Okafor C, Derfoul A, Danielson KG, Hall DJ, Tuan RS (2005) A three-dimensional nanofibrous scaffold for cartilage tissue engineering using human mesenchymal stem cells. Biomaterials 26:599
185. Thorvaldsson A, Stenhamre H, Gatenholm P, Walkenström P (2008) Electrospinning of highly porous scaffolds for cartilage regeneration. Biomacromolecules 9:1044
186. Sill TJ, von Recum HA (2008) Electrospinning: applications in drug delivery and tissue engineering. Biomaterials 29:1989–2006
187. Agarwal S, Wendorff JH, Greiner A (2008) Use of electrospinning technique for biomedical applications. Polymer 49:5603–5621
188. Zeng J, Yang L, Liang Q, Zhang X, Guan H, Xu X, Chen X, Jing X (2005) Influence of the drug compatibility with polymer solution on the release kinetics of electrospun fiber formulation. J Controlled Release 105:43
189. Xie J, Wang C-H (2006) Electrospun micro- and nanofibers for sustained delivery of paclitaxel to treat C6 glioma in vitro. Pharm Res 23:1817
190. Huang Z-M, He CL, Yang A, Zhang Y, Han X-J, Yin J (2006) Encapsulating drugs in biodegradable ultrafine fibers through co-axial electrospinning. J Biomed Mater Res Part A 77A:169
191. Cui W, Li X, Yu G, Zhou S, Weng J (2006) Investigation of drug release and matrix degradation of electrospun poly(D,L-lactide) fibers with paracetanol inoculation. Biomacromolecules 7:1623
192. Zeng J, Xu X, Chen X, Liang Q, Bian X, Yang L, Jing X (2003) Biodegradable electrospun fibers for drug delivery. J Controlled Release 92:227
193. Qi H, Hu P, Xu J, Wang A (2006) Encapsulation of drug reservoirs in fibers by emulsion electrospinning: morphology characterization and preliminary release assessment. Biomacromolecules 7:2327
194. Luong-Van E, Grondahl L, Chua KN, Leong KW, Nurcombe V, Cool SM (2006) Controlled release of heparin from poly(ε-caprolactone) electrospun fibers. Biomaterials 27:2042
195. Zhang Y, Wang X, Feng Y, Li J, Lim CT, Ramakrishna S (2006) Coaxial electrospinning of (fluorescein isothiocyanate-conjugated bovine serum albumin)-encapsulated poly(ε-caprolactone) nanofibers for sustained release. Biomacromolecules 7:1049
196. Casper CL, Yamaguchi N, Kiick KL, Rabolt JF (2005) Functionalizing electrospun fibers with biologically relevant macromolecules. Biomacromolecules 6:1998
197. Zeng J, Aigner A, Czubayko F, Kissel T, Wendorff JH, Greiner A (2005) Poly(vinyl alcohol) nanofibers by electrospinning as a protein delivery system and the retardation of enzyme release by additional polymer coatings. Biomacromolecules 6:1484
198. Richardson TP, Peters MC, Ennett AB, Mooney DJ (2001) Polymeric system for dual growth factor delivery. Nat Biotechnol 19:1029
199. Chew SY, Wen J, Yim EKF, Leong KW (2005) Sustained release of proteins from electrospun biodegradable fibers. Biomacromolecules 6:2017
200. Lu YK, Kim K, Hsiao BS, Chu B, Hadjiargyrou M (2003) Development of a nanostructured DNA delivery scaffold via electrospinning of PLGA and PLA-PEG block copolymers. J Controlled Release 89:341

201. Kretlow JD, Mikos AG (2008) From material to tissue: biomaterial development, scaffold fabrication, and tissue engineering. AlChE J 54:3048
202. Chen RR, Mooney DJ (2003) Polymeric growth factor delivery strategies for tissue engineering. Pharm Res 20:1103
203. Saltzman WM, Olbricht WL (2002) Building dug delivery into tissue engineering. Nat Rev Drug Discov 1:177
204. Hersel U, Dahmen C, Kessler H (2003) RGD modified polymers: biomaterials for stimulated cell adhesion and beyond. Biomaterials 24:4385

Chapter 2
Materials and Methods

2.1 Materials

Table 2.1 lists the polyesters used to fabricate scaffolds, specifying compositions, suppliers and polymer molecular weight distributions. It is pointed out that all copolymers employed possess a random distribution of their comonomers.

P(L)LA, PLA$_{75}$GA$_{25}$, P(LA-TMC), PCL and PEO were commercial polymers and they were used without further purifications.

P(D,L)LA and PLAGA copolymers, provided by the Institute of Polymers and Carbon Materials (Zabrze, Poland), were synthesized by ring opening polymerization using a zirconium-based initiator as previously described [1]. The use of these low toxicity initiators is particularly interested, as it was demonstrated that cell viability is higher when polymers are synthesized with zirconium compounds as catalysts instead of the widely used tin compounds [2]. These polymers were used after drying at 80 °C under vacuum in order to eliminate residual solvents employed during polymer purification steps.

Polymers supplied by the Centre for Biocatalysis and Bioprocessing of Macromolecules (New York, USA) (e.g. PPDL, P(PDL-CL) and P(PDL-DO)) were synthesized by ring opening polymerization catalyzed by Candida antartica Lipase B (CALB) as earlier described [3–5]. Enzyme catalyzed polymerizations allow to synthesize copolymers displaying a random distribution of monomeric units thanks to transesterification reactions promoted by CALB during the synthesis.

Chloroform (CLF), Dichloromethane (DCM), N,N-dimethylformamide (DMF), Dimethyl sulfoxide (DMSO), Methanol (MetOH), Tetrahydrofuran (THF), 2-Chloroethanol (CE), Acetone, 1,1,1,3,3,3-Hexafluoro-2-propanol (HFIP) and Ethanol (EtOH) were purchased by Sigma–Aldrich Co. and they were used without further purification.

The cationic macroinitiator for the Atom Transfer Radical Polymerization, Poly(2-(*N,N,N*-trimethylammonium iodide) ethyl methacrylate-*co*-bis-2,3-(2-bromoisobutyl) glycerol monomethacrylate) (MI; M$_n$ = 21.2 kg/mol, PDI = 1.34, by

C. Gualandi, *Porous Polymeric Bioresorbable Scaffolds for Tissue Engineering*,
Springer Theses, DOI: 10.1007/978-3-642-19272-2_2,
© Springer-Verlag Berlin Heidelberg 2011

Table 2.1 Polymers used for scaffold fabrication

Polymer	Composition (molar ratio)	Supplier	Molecular weight distribution
Poly(L)lactide (Lacea H.100-E) [P(L)LA]	–	Mitsui fine chemicals (Dusseldorf, Germany)	$M_w = 172$ kg/mol $M_w/M_n = 3.2$[a]
Poly(D,L)lactide [P(D,L)LA]	D:L = 50:50	Institute of polymers and carbon materials, polish academy of science (Zabrze, Poland)	$M_w = 155$ kg/mol $M_w/M_n = 2.4$[a]
Poly((L)lactide-co-glycolide) [PLA$_{90}$GA$_{10}$]	LA:GA = 90:10	Institute of polymers and carbon materials, polish academy of science (Zabrze, Poland)	$M_w = 20$ kg/mol $M_w/M_n = 2.1$[b]
Poly((D,L)lactide-co-glycolide) (Resomer RG 756 S) [PLA$_{75}$GA$_{25}$]	LA:GA = 75:25	Boehringer (Ingelheim, Germany)	$M_w = 170$ kg/mol $M_w/M_n = 2.0$[a]
Poly((D,L)lactide-co-glycolide) [PLA$_{65}$GA$_{35}$]	LA:GA = 65:35	Institute of polymers and carbon materials, polish academy of science (Zabrze, Poland)	$M_w = 41$ kg/mol $M_w/M_n = 2.1$[a]
Poly((D,L)lactide-co-glycolide) [PLA$_{50}$GA$_{50}$]	LA:GA = 50:50	Institute of polymers and carbon materials, polish academy of science (Zabrze, Poland)	$M_w = 81$ kg/mol $M_w/M_n = 2.4$[a]
Poly((L)lactide-co-trimethylene carbonate) (Resomer LT 706) [P(LA-TMC)]	LA:TMC = 70:30[c]	Boehringer (Ingelheim, Germany)	[d]
Poly(ε-caprolactone) (787 Tone) [PCL]	–	Union carbide Co. (New Jersey, USA)	$M_w = 74$ kg/mol $M_w/M_n = 2.3$[a]
Poly(ω-pentadecalactone) [PPDL]	–	Centre for biocatalysis and bioprocessing of macromolecules, polytechnic university (New York, USA)	$M_w = 128$ kg/mol $M_w/M_n = 2.0$[a]
Poly(ω-pentadecalactone-co-ε-caprolactone) [P(PDL-CL)]	PDL:CL = 69:31	Centre for biocatalysis and bioprocessing of macromolecules, polytechnic university (New York, USA)	$M_w = 240$ kg/mol $M_w/M_n = 8$[e]
Poly(ω-pentadecalactone-co-p-dioxanone) [P(PDL-DO)]	PDL:DO = 47:53	Centre for biocatalysis and bioprocessing of macromolecules, polytechnic university (New York, USA)	$M_w = 70$ kg/mol $M_w/M_n = 2.3$[e]

(continued)

Table 2.1 (continued)

Polymer	Composition (molar ratio)	Supplier	Molecular weight distribution
Poly(ethylene oxide) [PEO]	–	Sigma–Aldrich (Milan, Italy)	$M_w \sim 1000$ kg/mol[f]

[a] measured by gel permeation chromatography (GPC) in CLF at 25 °C by using polystyrene standards
[b] measured by GPC in THF at 25 °C by using polystyrene standards
[c] mass ratio
[d] supplier provides inherent viscosity = 1.4 ± 0.2 dl/g, measured in CLF 0.1% w/V at 25 °C
[e] measured by GPC in ortho-dichlorobenzene at 135 °C by using polystyrene standards
[f] provided by the supplier

GPC in DMF at 70 °C, by using polymethylmetacrilate standards), was synthesized as previously described [6]. Glycerol Monomethacrylate (GMMA; Cognis, Southampton, UK), 2,2'-Bipyridine (2,2'-Bpy; Sigma–Aldrich), CuCl and CuBr$_2$ (Sigma–Aldrich) were used without further purification.

A six-arms star-branched oligo(D,L)lactic acid (PLA-T6; $M_n = 25$ kg/mol by ^{1}H-NMR), synthesized as described by Biela et al. [7], was kindly provided by Prof. G. Di Silvestro (Organic and Industrial Chemistry Dept., University of Milan). In brief, the PLA-T6 was obtained by polycondensation reaction of lactide and the exa-functional initiator di-pentaerythritol (T6) catalyzed by Sn(Oct)$_2$. In order to transform the carboxyl end groups of PLA-T6 into carboxylate terminal groups, the oligomer was subject to the following salification procedure prior to use: 400 mg of PLA-T6 were dissolved in 20 ml of THF with the addition of 0.6 ml NaOH 0.1 M. The solution was stirred for 2 h then oligomer was precipitated in cyclohexane, washed with deionized water and dried over P$_2$O$_5$ under vacuum for 2–3 days.

2.2 Scaffold Fabrication by ScCO$_2$ Foaming

ScCO$_2$ scaffold fabrication was carried out in a 60 mL stainless steel high-pressure autoclave (made in house) (Fig. 2.1a) connected with a high pressure PM101 pump (New Ways of Analytics, Lörrach, Germany) that was used to charge CO$_2$ into the autoclave. Temperature and CO$_2$ pressure inside the autoclave were accurately controlled during the foaming process by using: (1) a CAL 3300 temperature controller (Advanced Industrial Systems, Inc., Luisville, USA) connected to a thermocouple inserted into the autoclave (Fig. 2.1b) and (2) a backpressure regulator (Bronkhorst, the Netherlands) and a pressure transducer (Fig. 2.1c).

A polymer disc (200 ± 5 mg) inserted in a cylindrical Teflon mould (10 mm diameter and 10 mm height, Fig. 2.1d) was placed in the autoclave and the foaming process was carried out as follows: the autoclave was heated to the desired temperature and filled with CO$_2$ at 230 bar (pressurization stage). The system was maintained at constant temperature and pressure over a given period of time (soak time). The soak stage was followed by a depressurization stage during

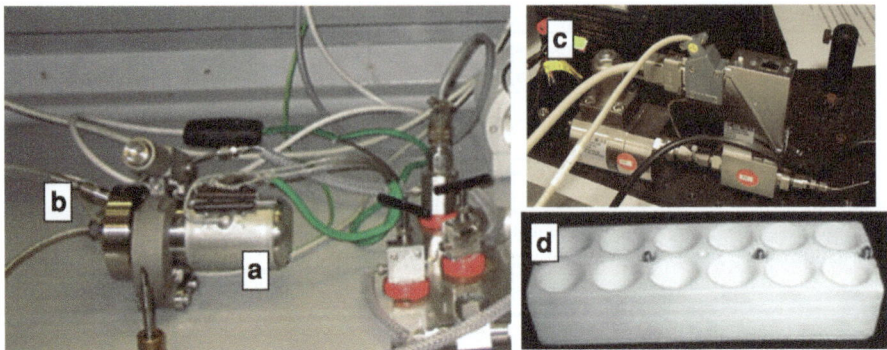

Fig. 2.1 ScCO$_2$ foaming apparatus **a** autoclave, **b** thermocouple, **c** back pressure regulator system and **d** Teflon mould with 12 wells for batch scaffold production. This mould was designed with a detachable base to allow easy removal of scaffolds after fabrication

Fig. 2.2 View cell for real-time foaming process observations

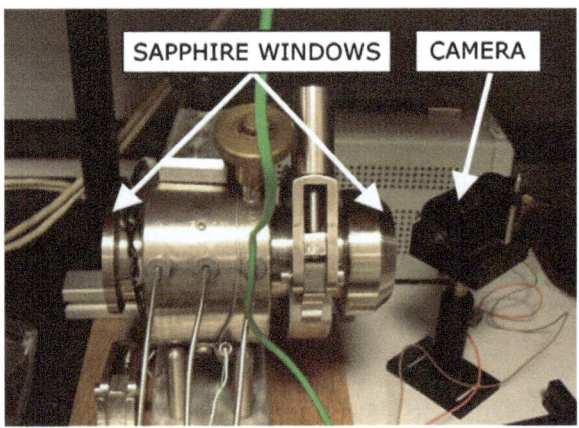

which the pressure was decreased to ambient pressure at controlled depressurization rate (dP/dt). During this stage the temperature was either kept constant or lowered at a controlled cooling rate (dT/dt) down to a selected temperature. At the end of the process, after a spontaneous cooling to room temperature (RT), the Teflon mould containing the foamed sample was removed from the autoclave.

Alternatively, the foaming process was also carried out in a 100 ml stainless steel high-pressure autoclave equipped with two sapphire windows (view cell, Fig. 2.2) with the aim to visualize the macroscopic changes of sample aspect and shape during the foaming process. Sapphire windows were located at each end of the autoclave, one used for back illumination. A polymer disc (200 ± 5 mg) was placed in a Teflon mould (10 × 10 × 2 mm) inserted in the view cell, and foaming process was carried out as previously described. A CCD uEye camera (Firstsight Vision, UK) placed in front of the sapphire window was used to capture real time images of the sample subjected to the foaming process.

2.3 Scaffold Fabrication by Electrospinning

The electrospinning (ES) apparatus was placed in a glove box (Iteco Eng.,
Ravenna, Italy, 100 × 75 × 100 cm) equipped with a temperature and humidity
control system (Fig. 2.3b). The ES apparatus (made in house) was composed of a
SL 50 p 10/CE/230 high voltage power supplier (Spellman, New York, USA,
Fig. 2.3c), a KDS-200 syringe pump (KDScientific Inc., Massachusetts, USA,
Fig. 2.3d), a glass syringe containing the polymer solution, a stainless-steel blunt-
ended N-P3-G18 needle (Hamilton, Bonaduz, Switzerland, Fig. 2.3e) connected
with the power supply electrode and a grounded collector (Fig. 2.3f). The polymer

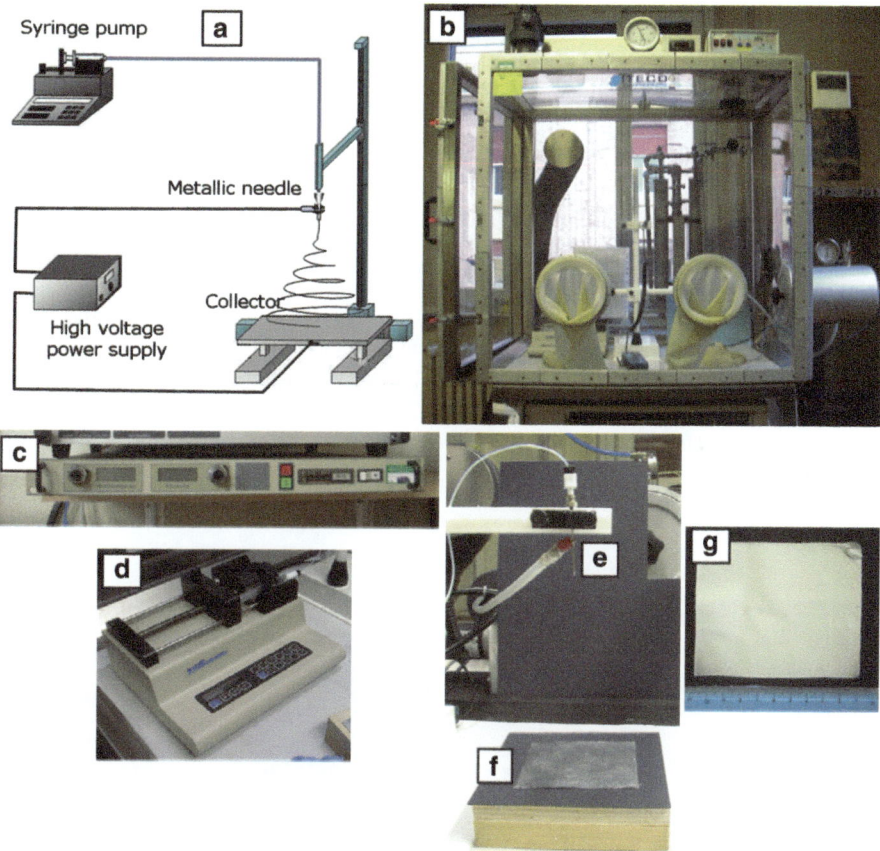

Fig. 2.3 a Scheme of the ES process, **b** glove box containing the ES apparatus composed of:
c high voltage power supply, **d** syringe pump, **e** metallic needle and **f** collector. **g** representative
picture of ES mat deposited on aluminium plate collector (10 × 10 mm)

solution was dispensed through a Teflon tube to the needle that was vertically placed on the target.

According to productivity and fibre deposition distribution requirements, collectors of different type and size were used. Alluminium plate collectors were employed for fabricating non-woven ES mats composed of randomly oriented fibres. Cylindrical rotating targets of different radius were used to collect ES fibres with different degree of spatial orientation. Finally, ad hoc developed targets, that allow to accurately control fibre deposition, were employed in order to fabricate patterned ES mats. Such collectors and the effect of their composition and geometry on mat morphology will be described in detail in Chap. 3.

ES polymer mats loaded with additives (i.e. Endothelial Cell Growth Factor Supplement and PLA-T6 oligomers) were obtained by simply electrospinning the polymeric solution containing the additional substance at the desired concentration.

In all cases, after fabrication, ES mats (Fig. 2.3g) were kept under vacuum over P_2O_5 at RT overnight in order to eliminate residual solvents.

2.3.1 Surface Modification

P(L)LA ES samples (3 ± 1 mg) containing 10% w/w of PLA-T6 oligomers were fixed on plastic rings (CellCrown^{TM}12, inner diameter = 15 mm, Scaffdex, Tampere, Finland, Fig. 2.4) and were immersed in EtOH for 15 min in order to ensure a fast and complete wetting of the intrinsically hydrophobic scaffold. EtOH was then replaced by deionized water through repeated rinses.

Each wet mat was placed in 10 ml of 0.1% w/V aqueous solution of the ATRP-macroinitiator (MI) and left at RT overnight under shaking to allow electrostatic adsorption to occur. Then, mats were thoroughly rinsed with deionized water and dried under nitrogen purge.

Fig. 2.4 ES mat fixed on a CellCrown^{TM}12 plastic ring (Scaffdex)

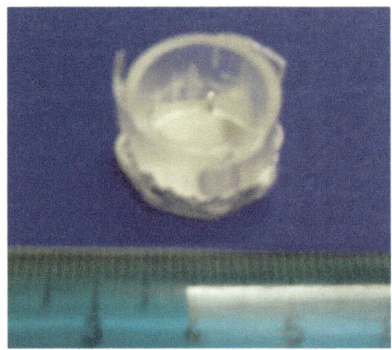

The obtained MI-coated mats were inserted in 50 ml Falcon tubes, placed in a parallel reactor and purged with nitrogen. Deionised water containing a mixture of GMMA, CuCl, CuBr$_2$ and 2,2'-Bpy (molar ratio = 60:1:0.3:2.8, GMMA concentration = 2.1 M) was bubbled with nitrogen for 45 min before addition of the required volume of MetOH (H$_2$O:MetOH = 1:1, by volume). Aliquots of this reactive mixture (22 ml) were then transferred to each Falcon tube containing ES mats to start polymerisation. The Surface-Initiated ATRP (SI-ATRP) of GMMA on nanofibers was carried out under nitrogen at RT for 16 h. The polymerization was interrupted by exposing mats to air. All samples were thoroughly washed with deionised water for one day.

2.3.2 In Vitro Degradation Experiments

Hydrolytic degradation studies were carried out on P(L)LA non-woven ES mats (25 ± 5 mg). Prior to degradation experiments specimens were dried over P$_2$O$_5$ under vacuum at RT for 2 days and they were weighted to yield the sample initial weight (m_0). Subsequently, samples were pre-wetted in EtOH for 15 min. EtOH was then replaced by deionized water through repeated rinses. Wet ES samples were immersed in phosphate buffered solution (0.1 M, pH = 7.4) and incubated in a shaking bath (SBS30 Stuart Scientific, Surrey, UK) at 37 °C and 50 revs/min. The buffer solution was periodically changed to keep the pH constant during the entire time scale of the degradation experiments. After selected exposure times, samples were recovered, repeatedly washed with deionized water to remove the buffer salt components and then dried over P$_2$O$_5$ under vacuum for 2 days to constant weight (m_x). The percentage weight remaining m(%) after buffer exposure was calculated according to Eq. 2.1:

$$m\,(\%\,) = 100 - \frac{m_0 - m_x}{m_0} \times 100 \qquad (2.1)$$

Where m_0 is sample initial weight and m_x is sample weight after x days in buffer at 37 °C.

2.3.3 Scaffold Preparation for Cell Culture Experiments

Scaffold fixation on plastic rings (Fig. 2.5a, b) was adopted in order not only to avoid cell dispersion/outflow during cell culture experiments, but also to improve scaffold handling and to prevent scaffold shrinkage during the subsequent cell culture steps. Cell culture experiments were performed by inserting the ES samples, preliminarily mounted on plastic rings, into common TCPS culture wells (Fig. 2.5c). During the seeding step, cells were confined onto the upper scaffold surface by the walls of the ring. By this means, cell migration towards the TCPS well bottom was prevented (Fig. 2.5d).

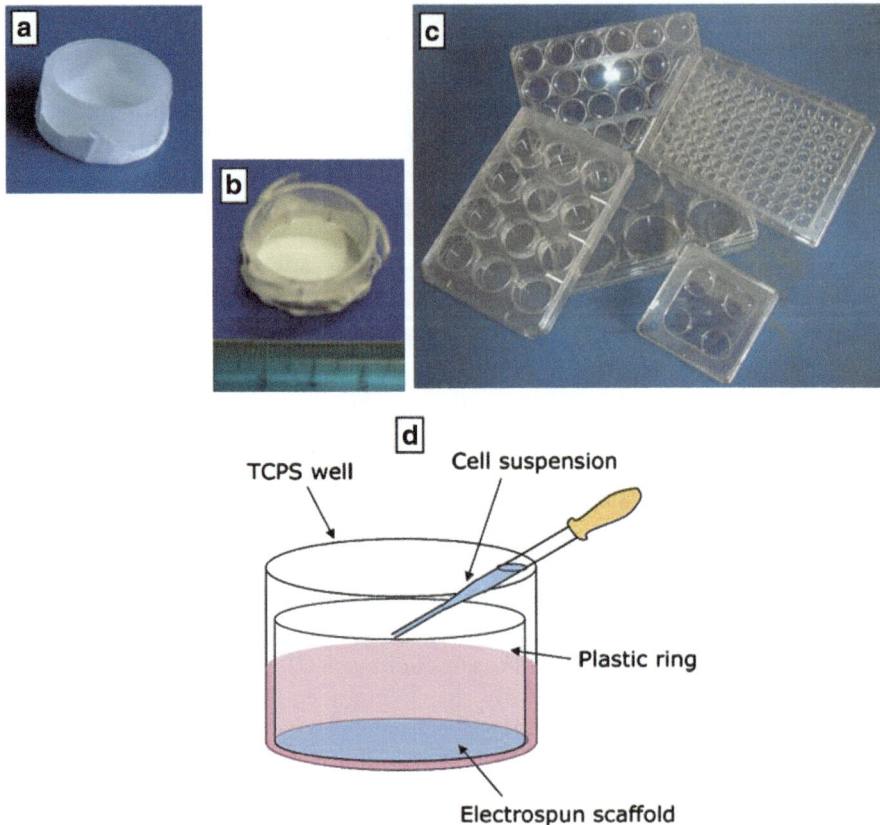

TCPS well   Cell suspension

Plastic ring

Electrospun scaffold

Fig. 2.5 a ES scaffold fixed on a Tecaflon plastic ring by using medical-grade silicon, **b** ES scaffold fixed on a CellCrown^TM6 plastic ring (Scaffdex), **c** common TCPS multiwell culture plates (different well dimensions are available) and **d** schematic representation of cell culture experiments: the ES sample fixed on the plastic ring is inserted into the TCPS well and cells are seeded on the upper surface of the scaffold

When cell culture experiments were performed in 12-multiwell TCPS plates (circular wells of 19 mm in diameter), ES scaffolds were fixed on Tecaflon (PVDF) plastic rings (internal diameter = 17 mm, external diameter = 18 mm) using silicone (GE Silicones Rubber, RTV 108Q, Fig. 2.5a). When cell culture experiments were performed in 6-multiwell TCPS plates (circular wells of 32 mm in diameter), CellCrown^TM6 plastic rings (inner diameter = 29 mm, Scaffdex, Tampere, Finland, Fig. 2.5) were used to fix the ES scaffolds without the need to use silicon glue.

Before exposure to culture medium, all scaffolds were subjected to a sterilization procedure using EtOH according to the following protocol: under a laminar flow, the scaffolds were immersed in 85% V/V EtOH for 15 min, followed by 70% V/V EtOH for 15 min, and then washed 3 times with phosphate buffered saline

(PBS, pH = 7.4) plus 2% Penicillin/Streptomycin (BioWhittaker-Lonza) and 0.2% Amphotericyn B (Sigma). Scaffolds were kept in this solution overnight under ultraviolet irradiation (TUV 30 W/G30 T8).

2.4 Characterization Methods

2.4.1 Thermogravimetric Analysis

Thermogravimetric analysis (TGA) measurements were carried out using a TGA2950 thermogravimetric analyzer (TA Instruments, New Castle, Delaware, USA). Analysis were performed on samples weighing 2–8 mg, from RT to 600 °C, at a heating rate of 10 °C/min, under N_2 flow.

2.4.2 Differential Scanning Calorimetry

Differential scanning calorimetry (DSC) measurements were carried out in helium atmosphere by using a Q100 DSC apparatus (TAInstruments, New Castle, Delaware, USA) equipped with a liquid nitrogen cooling system (LNCS) low-temperature accessory. Samples were placed in aluminum pans and subjected to heating scans at 20 °C/min from −80 °C to a temperature higher than glass transition temperature (T_g) for completely amorphous polymers, or higher than melting temperature (T_m) when semicrystalline polymers were analysed. Either quench cooling or controlled cooling at 10 °C/min were applied between heating scans. T_g values were taken at half-height of the glass transition heat capacity step while crystallization temperatures (T_c) and T_m were taken at the maximum of exotherm and endotherm peaks respectively. The degree of crystallinity, χ_c, was calculated using the following equation:

$$\chi_c = \frac{\Delta H_m}{\Delta H_m^0} \times 100 \tag{2.2}$$

Where ΔH_m is the experimental melting enthalpy obtained from the DSC scan and ΔH_m^0 is the melting enthalpy of 100% crystalline polymer.

2.4.3 Scanning Electron Microscopy

Samples were fixed with a conducting bi-adhesive tape on aluminium stubs and they were sputter coated with gold. Scanning electron microscopy (SEM) observations were carried out by using a Philips 515 microscope at an accelerating voltage of 15 kV. Images were acquired and analysed with EDAX Genesis software.

2.4.4 Micro X-Ray Computed Tomography

Micro X-ray Computed Tomography (μ-CT) images were acquired using a Skyscan 1174 Scanner (Skyscan, Aartselaar, Belgium). The scanner was set to a voltage of 50 kV and a current of 800 mA. By keeping constant the threshold range, the resulting 2D images were elaborated to obtain 3D reconstructions of the scaffolds, from which porosity and pore size were calculated, and pore interconnectivity was visually estimated.

2.4.5 Stress–Strain Analysis

Mechanical properties of foamed scaffolds were evaluated on 5 mm × 5 mm × 3 mm (thickness) specimens. Compression stress–strain measurements were performed with a TA.HDplus Texture Analyzer (Stable Micro Systems Ltd., Surrey, United Kingdom) at RT and at a cross head speed of 0.01 mm/s (load cell 750 N). Triplicate measurements were performed and average values (±standard deviation) are reported.

2.4.6 Wide Angle X-Ray Diffraction

Wide angle X-ray diffraction (WAXS) measurements were carried out at RT with a X'Pert PRO diffractometer (PANalytical, Almelo, the Netherlands) equipped with an XCelerator detector. Cu anode was used as X-ray source (K radiation at $\lambda = 0.15406$ nm, 40 kV, 40 mA) and 1/4 divergence slit was used to collect data in the range $2\theta = 2$–$60°$. After subtracting the diffractogram of an empty sample holder from the experimental diffraction curve, the amorphous and crystalline contributions were calculated by fitting method using the WinFit program. The degree of crystallinity (χ_c) was evaluated as the ratio of the crystalline peak areas to the total area under the scattering curve [8].

2.4.7 Gel Permeation Chromatography

Sample molar mass was evaluated by gel permeation chromatography (GPC) in chloroform (flow rate = 1 ml/min) at 35 °C by using a VE3580 solvent delivery system (Viscotek Corp., Texas, USA) with a set of two PLgel Mixed-C columns and a Shodex SE 61 refractive index detector. A volume of 100 μL of sample solution in chloroform (5% w/V) was injected. Polystyrene standards were used to generate a calibration curve.

2.4.8 ζ-Potential

Electrokinetic analyses were performed with a SurPASS electrokinetic analyzer (Anton Paar, Österreich, Austria) equipped with a cylindrical glass cell. ES samples pre-wetted in EtOH and thoroughly rinsed with deionized water were analysed. The wet sample was inserted into the cylindrical cell. The ζ-potential was determined from the measurement of streaming potential generated by the imposed movement of an electrolyte solution (KCl 1×10^{-3} M) through the sample. The ζ-potential, which is related to the charge density on sample surface, was determined at pH values in the range 5–9 by performing automatic titration.

References

1. Dobrzynski P, Kasperczyk J, Janeczek H, Bero M (2001) Synthesis of biodegradable copolymers with the use of low toxic zirconium compounds. 1. Copolymerization of glycolide with L-lactide initiated by Zr(Acac)$_4$. Macromolecules 34:5090
2. Czajkowska B, Dobrzynski P, Bero M (2005) Interaction of cells with L-lactide/glycolide copolymers synthesized with the use of tin or zirconium compounds. J Biomed Mater Res Part A 74A:591
3. Bisht KS, Henderson LA, Gross RA (1997) Enzyme-catalyzed ting-opening polymerization of ω-pentadecalactone. Macromolecules 30:2705
4. Ceccorulli G, Scandola M, Kumar A, Kalra B, Gross RA (2005) Cocrystallization of random copolymers of ω-pentadecalactone and ε-caprolactone synthesized by lipase catalysis. Biomacromolecules 6:902
5. Jiang Z, Azim H, Gross RA, Focarete ML, Scandola M (2007) Lipase-catalyzed copolymerization of ω-pentadecalactone with p-dioxanone and characterization of copolymer thermal and crystalline properties. Biomacromolecules 8:2262
6. Edmondson S, Vo CD, Armes SP, Unali GF (2007) Surface polymerization from planar surfaces by atom transfer radical polymerization using polyelectrolytic macroinitiators. Macromolecules 40:5271
7. Biela T, Duda A, Penczek S, Rode K, Pasch H (2002) Well-defined star polylactides and thier behaviour in two-dimensional chromatography. J Polym Sci Part A Polym Chem 40:2884
8. Kakudo M, Kasai N (1972) X-ray diffraction by polymers. American Elsevier Publishing, New York

Chapter 3
Results and Discussion

The present chapter is divided into four main sections:

1. Porous scaffold fabrication by supercritical carbon dioxide ($scCO_2$) foaming;
2. Porous scaffold fabrication by electrospinning (ES);
3. In vitro hydrolytic degradation of ES scaffolds;
4. Cell culture experiments.

The first two sections illustrate scaffold fabrication technologies employed in the present research (i.e. $scCO_2$ foaming and electrospinning) and describe porous scaffold production and characterization. Experiments of $scCO_2$ foaming were carried out at the Inorganic and Material Chemistry Department and at the Centre of Biomolecular Science (University of Nottingham). ES technology was implemented in the course of the research thanks to the collaboration with Mechanical Engineering and Electrical Engineering Departments (University of Bologna). Finally, the preliminary experiments of ES scaffold surface functionalization, illustrated at the end of the second section, were carried out at the Laboratory of Polymers and Biomaterials (University of Manchester).

The third section reports hydrolytic degradation experiments on poly(L)lactide ES scaffold whereas the fourth section illustrates cell culture experiments performed by using fibrous scaffolds. The latter were carried out with the collaboration of the Biochemistry Department "G. Moruzzi" (University of Bologna), the Clinical Department of Radiological and Histocytopathological Sciences (University of Bologna) and the Foundation for Development of Cardiac Surgery (Poland).

The two techniques employed to fabricate scaffolds (i.e. $scCO_2$ foaming and ES) are based on different approaches. The first one uses CO_2 as porogen that is subsequently removed from the polymer matrix, leaving behind a porous structure. The second one produces a fibre that, thanks to its continuous deposition, generates a non-woven porous structure. The obtained scaffolds are completely different in terms of macroscopic aspect, of 3D microscopic morphology and of mechanical properties. Figure 3.1 shows pictures of both a foamed scaffold fabricated by $scCO_2$ foaming

C. Gualandi, *Porous Polymeric Bioresorbable Scaffolds for Tissue Engineering*, 43
Springer Theses, DOI: 10.1007/978-3-642-19272-2_3,
© Springer-Verlag Berlin Heidelberg 2011

Fig. 3.1 Pictures and SEM micrographs of representatives **a** foamed scaffold and **b** ES scaffold

Fig. 3.1 Pictures and SEM micrographs of representatives **a** foamed scaffold and **b** ES scaffold

and of a non-woven mat produced by ES (Fig. 3.1a, b, respectively). The foam is a 3D object that assumes the shape of the container where it is produced (a cylindrical mould in this case), whereas the ES scaffold is a flexible sheet less then 1 mm thick. SEM micrographs show the microscopic morphology of foamed scaffold compared with that of ES mat: the first one is characterized by large circular pores of hundreds of microns whereas the second one has much smaller pores, whose dimension is strictly related to fibre diameter. As a matter of fact, sub-micrometric fibres generate pores of few microns whereas micrometric fibres lead to pores in the range 10–100 μm [1].

Since mechanical performance is strictly related to the microstructure, the two kinds of scaffold can be employed for different applications. Foamed scaffolds can assume a structural supporting role when used as tissue replacements and they are, therefore, particularly suitable in hard TE (e.g. bone or cartilage TE, depending on the type of material). On the contrary, ES mats are flexible sheets that deform easily under bending deformations, being appropriate for replacement of soft tissues such as cardiac, vascular or nervous tissues. Given the intrinsic biomimetic features of ES materials they are considered the most promising scaffolds in TE. Therefore, the present research activity was mainly dedicated to study and to develop ES process and ES products. For the same reason, in the course of the present Ph.D., cell culture experiments were only carried out on ES scaffolds.

3.1 Porous Scaffold Fabrication by ScCO$_2$ Foaming

Foamed scaffolds have been typically produced to date using amorphous polymers, in particular non biodegradable polymethylmethacrylates [2–4] and biodegradable polyesters [5–9]. The classical foaming process exploits the capability of scCO$_2$ to plasticize a polymer glass. Typically, polymer foaming (Fig. 3.2) involves three stages: (1) the pressurization stage (Fig. 3.2 blue), (2) the soak stage (Fig. 3.2 red) and (3) the depressurization stage (Fig. 3.2 yellow). During the first stage the glassy polymer absorbs scCO$_2$ that, acting as plasticizer, lowers the T$_g$ and changes the polymer state from glassy to rubbery. The polymer is maintained at high constant pressure during the soak stage, where it continues to absorb scCO$_2$. During the subsequent depressurization stage, run at constant temperature, the pressure is reduced, CO$_2$ phase changes from supercritical to gas, and bubble nucleation and gas bubble growth occur in the rubbery polymer, generating pores. Concomitantly, as a consequence of plasticizer concentration reduction, the polymer T$_g$ increases and the porous structure is fixed in a solid glassy state [10–12]. Foam 3D structure strongly depends on process parameters such as soak time (t$_{soak}$), soak pressure (P$_{soak}$), depressurization rate (dP/dt) and process temperature (T$_{process}$).

ScCO$_2$ foaming is relatively simple to perform for completely amorphous polymers thanks to the large T$_g$ depression occurring in the presence of scCO$_2$.

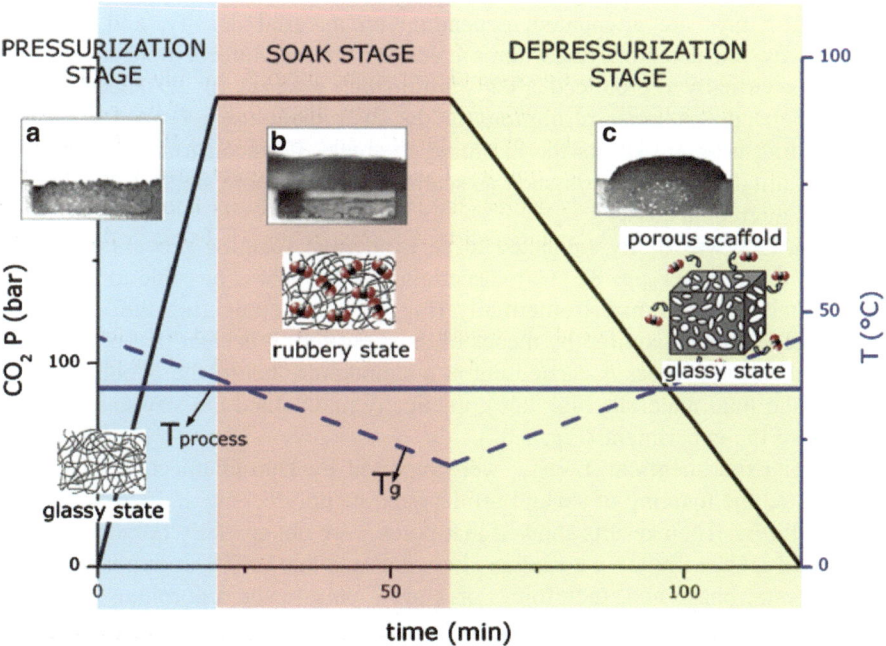

Fig. 3.2 Scheme of amorphous polymer foaming process. CO$_2$ pressure changes during the process (*black line*) whereas temperature is kept constant (*blue line*). The dotted blue line depicts the polymer glass transition temperature changes during the foaming process. Pictures (acquired with the view cell) show the behaviour of a representative amorphous polyester during the process: **a** initial granular polymer, **b** polymer-scCO$_2$ mixture in the soak stage and **c** porous scaffold at the end of the depressurization stage

On the contrary, the foaming process of semicrystalline polymers can be more complicated, owing to the very low CO_2 diffusivity in the polymer crystal phase [13, 14].

In the course of the present Ph.D., the above described foaming process was modified and adapted in order to foam a poly(ω-pentadecalactone-*co*-ε-caprolactone) (P(PDL-CL), CL = 31 mol%) [15]. This polymer is highly crystalline (crystallinity degree $\chi_c > 70\%$ by WAXS [16]) with $T_m = 82$ °C (by DSC). The peculiar crystallization behaviour of P(PDL-CL) copolymers has been recently investigated [16]. It was demonstrated that, as a result of cocrystallization of the PDL and CL units, these random copolymers possess a high degree of crystallinity over the whole composition range, with melting temperature that smoothly decreases with increasing CL content from that of PPDL ($T_m = 100$ °C) to that of PCL ($T_m = 60$ °C). Another interesting aspect is that copolymerization of the long hydrophobic PDL unit with the more hydrophilic CL monomer is expected to allow tuning of the hydrolytic degradation rate of these materials in view of their use as bioresorbable scaffolds.

Micro X-Ray Computer Tomography (μ-CT) was used to characterize foamed scaffold morphology. This non-destructive technique employs X-ray to obtain information on the interior part of the scaffold without sectioning it. During the scanning, X-rays are attenuated, depending on material density, and they are captured by the detector. Algorithms calculate attenuation coefficients and the sample is virtually reproduced after splitting into a series of 2D slices. Image quality depends on pixel resolution (in the instrument used 1 pixel = 6 μm). A modelling program stacks the 2D maps to create 3D reconstructions and provides quantitative information such as scaffold porosity, pore size, wall thickness and pore interconnectivity.

P(PDL-CL) was firstly subjected to an isothermal foaming process at a temperature lower than $T_m = 82$ °C, to ascertain whether it was possible to foam this semicrystalline copolymer isothermally (Fig. 3.3). By comparing the images of P(PDL-CL) sample acquired by means of the view cell, it is clear that no foaming occurs when the process is carried out at a temperature below the copolymer T_m, because the final material (Fig. 3.3c) practically maintained the size and shape it had before the experiment (Fig. 3.3a).

Similar experiments at $T < T_m$ were reported by Duroidiani et al. [17] who investigated the foaming of several semicrystalline polymers with different degree of crystallinity. Their results showed that pores were not spatially homogeneously distributed in the samples. The authors hypothesized that scCO$_2$ does not penetrate in the crystal phase and, therefore, pores grow only in the amorphous domains. Moreover, they found that for highly crystalline polymers no foaming at all occurred at $T < T_m$. Therefore, it is reasonable to conclude that, in the present case, given the low CO_2 diffusivity in the crystal phase, at $T < T_m$, there is a large amount of a solid non-plasticized phase that prevents bubble growth and pore formation in P(PDL-CL). This suggests that foaming process should be performed at a temperature higher than P(PDL-CL) T_m, in order to destroy the crystal phase

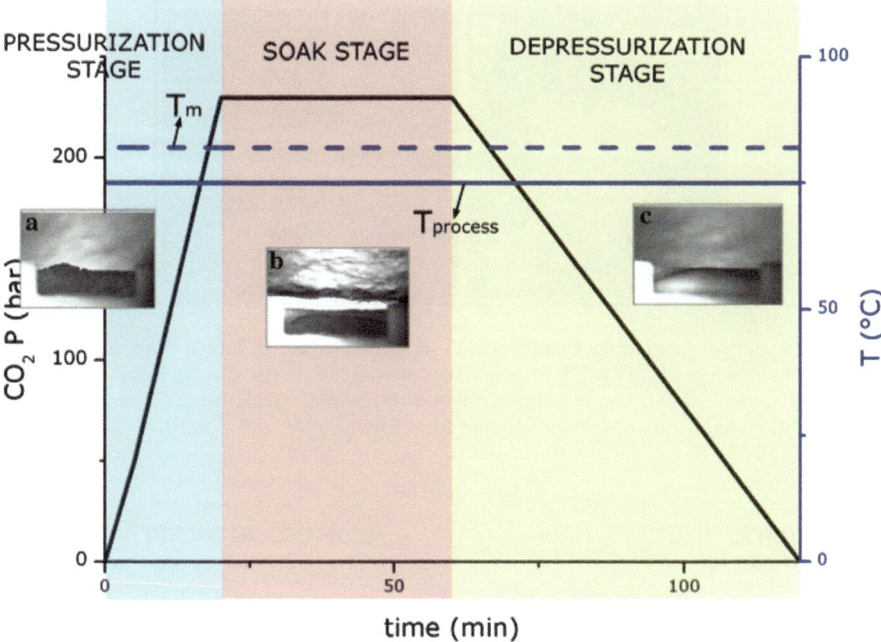

Fig. 3.3 Scheme of amorphous polymer foaming process applied to semicrystalline P(PDL-CL). CO$_2$ pressure changed during the process (*black line*) whereas temperature was kept constant (*blue line*). Pictures (acquired with the view cell) show that P(PDL-CL) sample did not foam during the process: **a** starting disc, **b** polymer-scCO$_2$ mixture and **c** not foamed P(PDL-CL) disc at the end of the depressurization stage

and to guarantee a homogeneous absorption of scCO$_2$ by the molten polymer during the soak stage.

Figure 3.4a shows a picture of P(PDL-CL) obtained after an isothermal foaming at T > T$_m$ together with a 2D μ-CT sample cross section perpendicular to the direction of foaming (Fig. 3.4b). It is clear that, when the process was performed at T > T$_m$, the sample underwent an increase in height, demonstrating that a foam formed. No increase in sample diameter occurred, since polymer foaming was confined in the radial direction by the walls of the mould. These experiments prove that CO$_2$ can be absorbed by the melt phase during the soak stage. However, the foamed sample had a hollow in the middle and a moderate number of very large pores (see Fig. 3.4b), as a consequence of structure collapse at a temperature that does not allow polymer solidification.

Hence, the foaming procedure was modified as follows. The experiments were carried out at a soak temperature, T$_{soak}$ > T$_m$, that was allowed to decrease during the depressurization stage to a value lower than polymer T$_m$ in order to promote crystallization and to fix the pores in a solid structure. This alternative foaming process (Fig. 3.5) applied to the highly crystalline P(PDL-CL) enables to obtain 3D porous foamed scaffolds from this polymer.

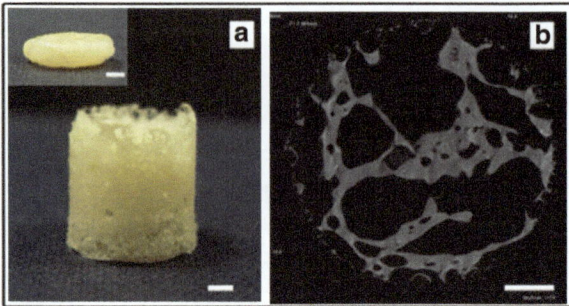

Fig. 3.4 Isothermal foaming at T = 90 °C (T_m = 82 °C, by DCS). Insert: starting P(PDL-CL) disc, **a** foamed sample **b** 2D μ-CT of a sample cross section perpendicular to the direction of foaming (P_{soak} = 230 bar, t_{soak} = 20 min, dP/dt = 8 bar/min). Scale bars = 2 mm (Reprinted from [15], Copyright (2010), with permission from Elsevier)

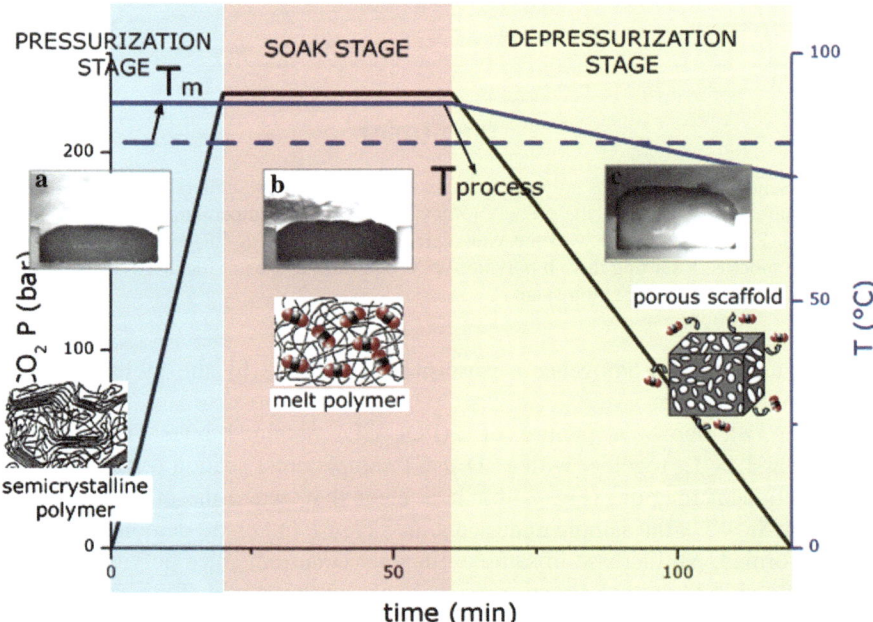

Fig. 3.5 Scheme of semicrystalline polymer foaming process. Both CO_2 pressure (*black line*) and temperature (*blue line*) change during the process. Pictures (acquired with the view cell) show changes of P(PDL-CL) sample shape during the process: **a** starting disc, **b** polymer-scCO$_2$ mixture and **c** porous scaffold at the end of the depressurization stage

An accurate control of foaming process parameters is necessary in order to fine tune scaffold 3D structure (e.g. porosity, pore size distribution and pore interconnectivity). Indeed, each parameter has a specific effect on final morphology. For instance, T_{soak}, P_{soak} and t_{soak} determine the amount of scCO$_2$ absorbed by the polymer melt during the soak stage. Subsequently, during the depressurization stage, nucleation and growth of CO$_2$ bubbles occurs. According to nucleation theory [11], the activation energy that must be overcome in order to create stable nuclei depends on pressure gradient, concentration gradient and temperature [18]. In the present experiments nucleation was forced to occur as a consequence of the pressure drop that decreased gas solubility and caused CO$_2$ supersaturation in the polymer. Nucleation density and bubble growth rate depend on depressurization rate (dP/dt) and on cooling rate (dT/dt). With the aim of understanding the effect of each variable on scaffold morphology, foaming experiments on the highly crystalline P(PDL-CL) copolymer were carried out by changing one process parameter at a time.

Figure 3.6 compares pictures of samples obtained by applying four different cooling rates (dT/dt) after soaking at 90 °C. 2D μ-CT reconstructions of sample cross sections perpendicular to the direction of foaming, 3D μ-CT reconstructions and pore size distributions are reported. Porosity (p) and average pore size (d) values are also given. Figure 3.6a and b show that low cooling rates generate large pores with a high degree of interconnectivity whereas high cooling rates lead to narrow distributions of small pores with few interconnections (Fig. 3.6c, d).

These results can be explained by considering that bubble growth, pore wall breaking and pore fixation are controlled by polymer viscosity, which in turn depends not only on temperature but also on the amount of CO$_2$ dissolved in the polymer [28]. The latter is regulated by the depressurization rate (dP/dt) which was kept constant in the present experiments. Therefore, in the applied conditions, the viscosity of the system essentially depends on cooling rate (dT/dt). At low cooling rates viscosity increases very slowly and pores grow uncontrolled because of the high diffusion of CO$_2$ through the highly mobile polymer chains (see Fig. 3.6a, b). On the contrary, rapid cooling rates lead to a fast increase of viscosity that inhibits bubble growth, leading to the formation of small closed pores (see Fig. 3.6d). As pointed out in Chap. 1, a key issue in the field of TE is the obtainment of scaffolds with interconnected pores. As regards P(PDL-CL) foamed scaffolds, acceptable pore interconnectivity was achieved by using intermediate cooling rates that promote formation of pore connections while preventing structure collapse. When 0.23 °C/min was applied a scaffold with interconnected pores with average diameter = 255 μm and 70% porosity (Fig. 3.6c) was fabricated. These experiments show that a very accurate control of the cooling rate is needed in order to carefully control the 3D structure. Similar conclusions were earlier reported in a study of scCO$_2$ foaming of P(L)LA, a much less crystalline polyester than P(PDL-CL) [19].

Figure 3.7 shows the effect of soak time on scaffold morphology. A non-homogeneous structure, characterized by a non porous core and a porous shell, was obtained with a short soak time (1 min, Fig. 3.7a). A more homogeneous scaffold with an average pore size of 155 μm and porosity of 57% was obtained by increasing the t$_{soak}$ to 10 min (Fig. 3.7b). A further increase of t$_{soak}$ (Fig. 3.7c) led to a decrease of average pore size down to 110 μm, whereas porosity remained almost unchanged. An analogous influence of soak time on scaffold structure was earlier reported by Goel et al. [11] and by Tai et al. [5] in foaming amorphous polymers. Therefore, the

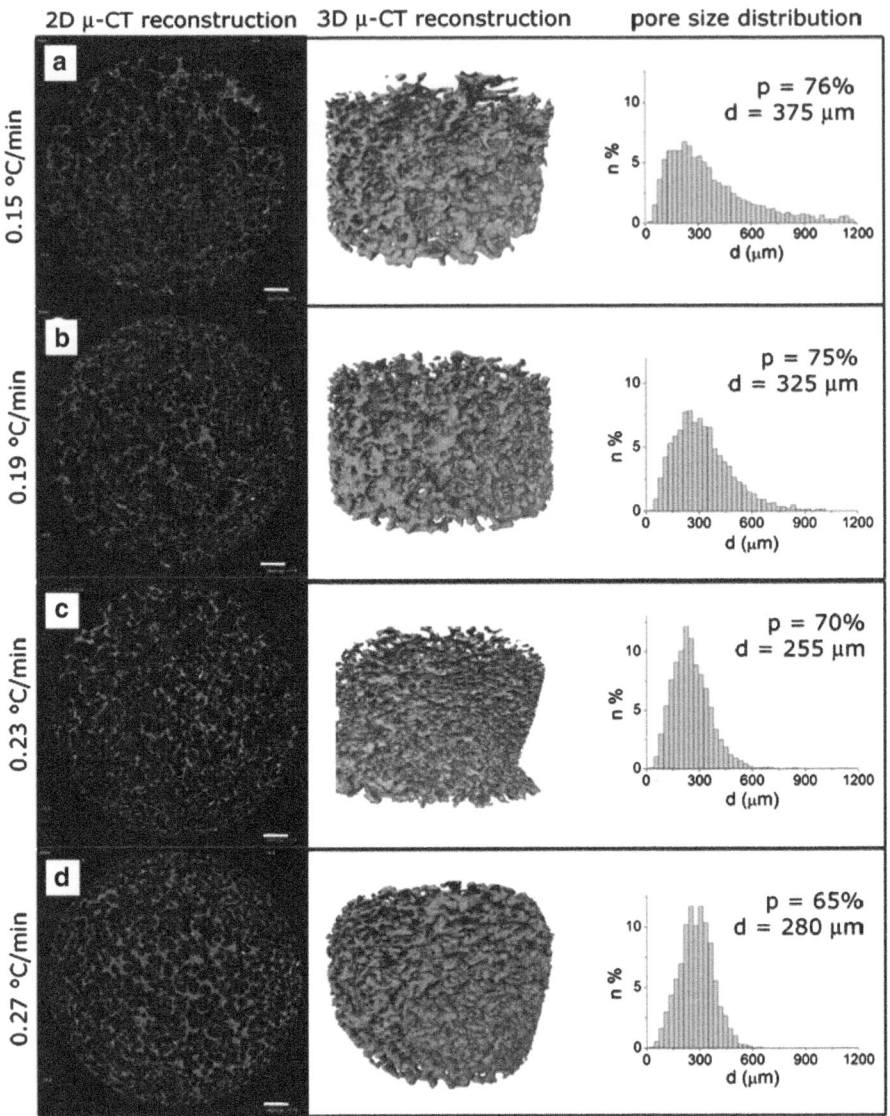

Fig. 3.6 Effect of cooling rate on scaffold morphology. P(PDL-CL) foamed at **a** dT/dt = 0.15 °C/min, **b** dT/dt = 0.19 °C/min, **c** dT/dt = 0.23 °C/min and **d** dT/dt = 0.27 °C/min (T_{soak} = 90 °C, P_{soak} = 230 bar, t_{soak} = 20 min, dP/dt = 4 bar/min). Scale bar = 1 mm (Reprinted from [15], Copyright (2010), with permission from Elsevier)

present results suggest that the t_{soak} parameter similarly affects the scaffold 3D structure of both semicrystalline and of amorphous polymers. Indeed, in both cases a polymer matrix, that is a melt in the case of P(PDL-CL) or a rubber when amorphous polymers are foamed, homogeneously absorbs $scCO_2$. The longer the soak time,

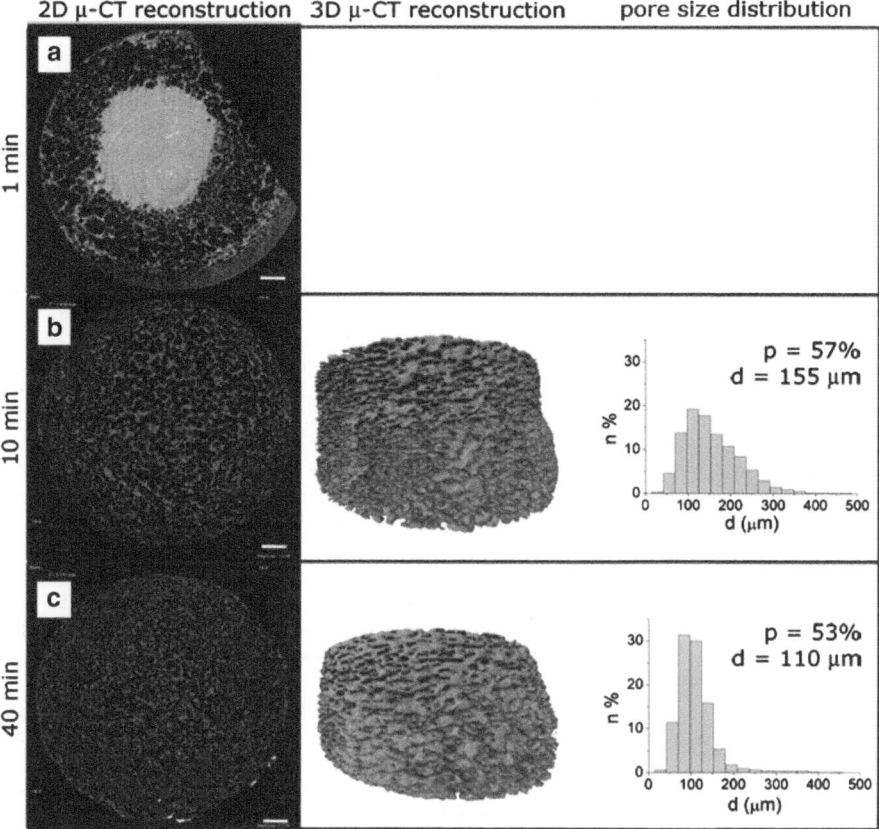

Fig. 3.7 Effect of soak time on scaffold morphology. P(PDL-CL) foamed at **a** t_{soak} = 1 min, **b** t_{soak} = 10 min and **c** t_{soak} = 40 min (T_{soak} = 90 °C, P_{soak} = 230 bar, dP/dt = 8 bar/min, dT/dt = 0.27 °C/min). Scale bar = 1 mm (Reprinted from [15], Copyright (2010), with permission from Elsevier)

the higher is the amount of CO_2 absorbed. High concentrations of CO_2 dispersed in the polymer generate a high density of nucleation sites during the depressurization stage, which lead to the fabrication of scaffolds with small pores. Conversely, larger pore size can be obtained by shortening the soak time. However the results of Fig. 3.7 show that t_{soak} cannot decrease indefinitely because non-homogenous 3D structures are obtained when the soak time is too short to allow the homogeneous diffusion of CO_2 throughout the polymer (see Fig. 3.7a).

After investigating the effect of cooling rate and soak time, the influence of the depressurization rate (dP/dt) on scaffold morphology was evaluated by keeping constant all other parameters. It is pointed out that experiments that differed in depressurization rate had to be run at different cooling rate, in order to operate over the same temperature range. Figure 3.8 shows that fast venting lead to

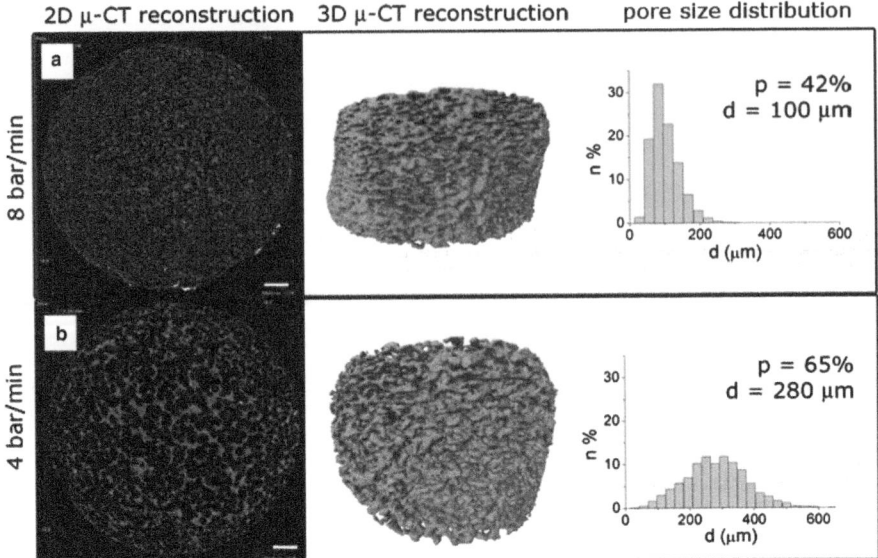

Fig. 3.8 Effect of depressurization rate on scaffold morphology. P(PDL-CL) foamed at **a** dP/dt = 8 bar/min and dT/dt = 0.54 °C/min and **b** dP/dt = 4 bar/min and dT/dt = 0.27 °C/min (T_{soak} = 90 °C, P_{soak} = 230 bar, t_{soak} = 20 min). Scale bar = 1 mm (Reprinted from [15], Copyright (2010), with permission from Elsevier)

fabrication of structures with small closed pores and low porosity (Fig. 3.8a) whereas decreasing the depressurization rate helps increasing average pore size, porosity and pore interconnections (Fig. 3.8b). Indeed, at lower depressurization rates pore growth lasts longer, thus generating larger and better interconnected pores. The present highly crystalline and earlier investigated amorphous polymers [3, 5] show similar dependence of scaffold morphology on depressurization rate. Indeed, in both cases, CO_2 bubbles grow and expand within a polymer matrix while it is undergoing solidification. Solidification of amorphous polymers occurs spontaneously during depressurization, where CO_2 evaporates, thus plasticizer concentration in the polymer decreases with a consequent increase of T_g. Differently to amorphous polymers, P(PDL-CL) solidifies owing to the decrease of temperature during the process, which induce polymer crystallization to occur.

The above described experiments have demonstrated that scCO_2 foaming is a versatile technique that can be applied to fabricate porous scaffolds not only from amorphous polymers but also from polymers possessing high crystallinity degree. However, working with semicrystalline polymers makes the foaming procedure difficult to control both because the process is non-isothermal and because solidification implies not only amorphous phase vitrification but also crystallization. Crystallization kinetics depend both on CO_2 concentration and temperature. The plasticizer decreases both T_g [20–22] and T_m [23], with T_g decreasing faster than T_m as a function of plasticizer concentration. This implies that the polymer

Fig. 3.9 Representative
DSC first scan of P(PDL-CL)
foam (heating rate =
20 °C/min) (Reprinted from
[15], Copyright (2010), with
permission from Elsevier)

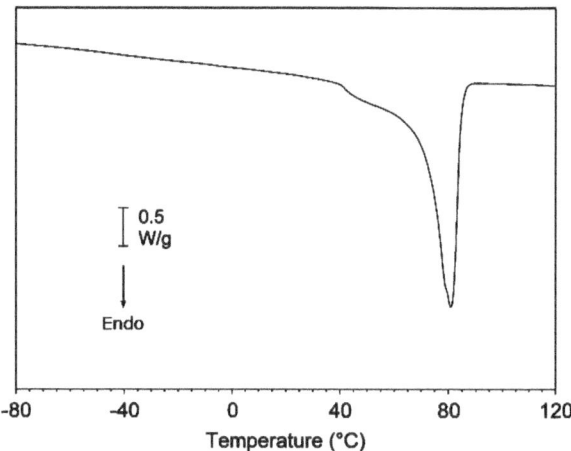

crystallization window, i.e. the range between T_g and T_m, shifts towards lower temperature and changes its breadth.

With the aim to analyze possible differences of scaffold crystallinity induced by changing foaming conditions, the thermal properties of the foamed copolymer were investigated by DSC. DSC first scans of the starting copolymer and of scaffolds differently foamed were compared. Surprisingly, all DSC curves were identical (see Fig. 3.9 as an example) with a single melting peak at 82 ± 1 °C and an associated melting enthalpy of 105 ± 2 J/g, whereas the glass transition heat capacity step was not detectable owing to the presence of a high amount of crystal phase. This result shows that crystal phase development in P(PDL-CL) foams is not affected by cooling rate, depressurization rate and soak time, within the range investigated. This observation can be attributed to the peculiar crystallization behaviour of P(PDL-CL) copolymers, where PDL and CL co-units crystallize in the same lattice with very high crystallization kinetics, thus developing the same amount of crystal phase regardless of the foaming process conditions. It is pointed out that the same behaviour is not expected for slowly crystallizable polymers, such as P(L)LA, for which foaming conditions can be extremely important in controlling the development of the crystal phase.

The mechanical properties of P(PDL-CL) scaffolds were characterized by compression tests. Compression experiments on P(PDL-CL) scaffolds yielded stress–strain curves (see Fig. 3.10 as an example) typical of porous materials [24], that can be described as follows: (1) an initial linear elastic region controlled by pore walls bending, (2) a stress plateau due to pore collapse and (3) a stress increasing region (densification region) where pores are completely collapsed and the load is applied to a bulk-like material.

Since the above discussed DSC results showed that the ratio of crystal to amorphous phase was unaffected by the foaming procedure, differences in mechanical properties must be attributed to scaffold 3-D structure. Table 3.1 lists the compressive mechanical test results of scaffolds obtained in different foaming

Fig. 3.10 Representative stress–strain compression curve of P(PDL-CL) foam. Three regions can be distinguished: (i) a linear elastic region, (ii) a stress plateau and (iii) a densification region

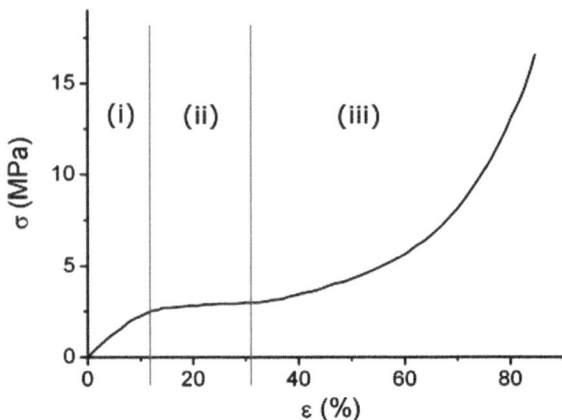

Table 3.1 Mechanical properties (from compression stress–strain curves) of selected scaffolds

Foaming conditions	Porosity[a]	Average pore size[a]	E (MPa)	σ_y (MPa)
$T_{soak} = 90$ °C, $P_{soak} = 230$ bar, $t_{soak} = 20$ min, $dP/dt = 4$ bar/min, $dT/dt = 0.38$ °C/min	51 ± 1	186 ± 13	21 ± 2	1.82 ± 0.07
$T_{soak} = 90$ °C, $P_{soak} = 230$ bar, $t_{soak} = 20$ min, $dP/dt = 4$ bar/min, $dT/dt = 0.27$ °C/bar	61 ± 5	206 ± 7	12 ± 6	1.4 ± 0.4
$T_{soak} = 90$ °C, $P_{soak} = 230$ bar, $t_{soak} = 20$ min, $dP/dt = 4$ bar/min, $dT/dt = 0.23$ °C/bar	72 ± 2	272 ± 50	4 ± 1	0.44 ± 0.06

Reprinted from [15], Copyright (2010), with permission from Elsevier
[a] from μ-CT analysis

conditions together with intrinsic scaffold parameters (porosity and pore size). Both compressive modulus (E) and compressive strength at yield (σ_y) decrease with increasing porosity and pore dimension. In particular, compressive modulus values are in the same range as that of articular cartilage [25, 26]. Moreover, Fig. 3.10 shows that P(PDL-CL) foams display a linear elastic deformation that extended up to 10% strain. This preliminary study of scaffold mechanical properties suggests that P(PDL-CL) foams may find applications as cartilage substitutes.

3.2 Porous Scaffold Fabrication by Electrospinning

By means of a very simple and cost-effective experimental apparatus, ES technology enables the production of polymeric nanofibres by establishing a high voltage difference between a metallic needle ejecting the polymeric solution and a collector. The technique involves an extremely complex electro-fluidodynamical process that has been the subject of intensive research aimed at modelling the physics of electrically driven jets and the evolution of jet instability [27–31]. A brief and simplified description of the process can be provided referring to

Reneker's separation of the ES process into four key stages: (1) launching of the jet, (2) elongation of the jet straight segment, (3) development of whipping instability and (4) fibre solidification and collection [32] (Fig. 3.11). In the absence of an applied electrical field, the droplet of solution coming out from the needle falls under the influence of gravity. When the solution is positively charged, the droplet changes its shape from spherical to elliptical. Once a certain threshold voltage is reached, the drop assumes a conical shape, generating the so called Taylor's cone—a term intending to honour Taylor's contribution to the understanding of droplet behaviour in electrical fields [33]—from which a jet emerges. Jet formation occurs when electrostatic force overcomes the surface tension and tends to deform the droplet in order to increase the surface area and minimize charge repulsion (Fig. 3.11i). The same interplay between surface tension and electrostatic force induces a straight elongation of the jet in the direction of the collector. The presence of entanglements that physically link the macromolecular chains allows the formation of a continuous jet (Fig. 3.11ii). During its travel towards the collector, the jet becomes unstable and displays bending and whipping movements. Its rapid and symmetric motion generates a cone-shaped envelope. At this stage the jet reduces its diameter and its surface area increases dramatically (Fig. 3.11iii). At the end of this stage all the solvent has evaporated and a continuous single fibre lays randomly oriented onto the collector (Fig. 3.11iv).

Morphology of the collected fibres depends on a number of variables, that are commonly classified into three groups:

A. solution parameters (e.g. concentration, electrical conductivity, dielectric constant, viscosity and vapour pressure of the polymer solution, etc.);
B. instrumental parameters (e.g. applied voltage (ΔV), needle to collector distance (d), solution flow rate (R), etc.);
C. environmental parameters (e.g. temperature (T), relative humidity (RH), etc.).

Fig. 3.11 Scheme of key stages of ES process according to Reneker et al. [32]

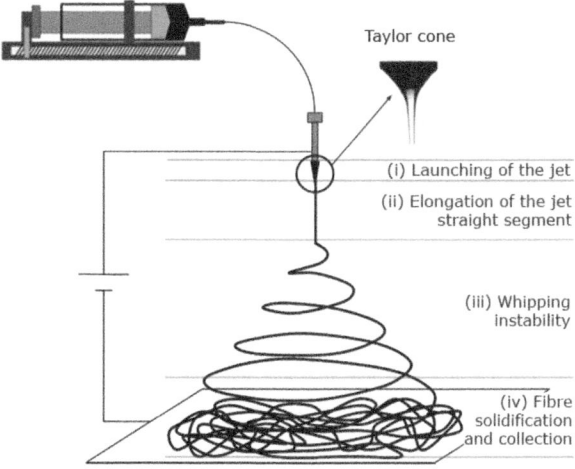

Taylor cone

(i) Launching of the jet

(ii) Elongation of the jet straight segment

(iii) Whipping instability

(iv) Fibre solidification and collection

None of the above mentioned variables plays an independent role on the final fibre morphology. Indeed, a given parameter often differently affects fibre morphology, depending on the specific polymer–solvent combination and on process variables. For instance, the effect of electrical potential difference ΔV on fibre diameter depends on the ES systems considered: ΔV can either have no effect on fibre diameter [34, 35], or its increase can decrease fibre diameter [36, 37] or it can change diameter distribution width [38, 39]. The understanding of how ES variables interplay in affecting fibre morphology is far from being understood and no predictive models have been developed. It turns out that suitable ES process conditions for the obtainment of the desired fibre morphology are always found by means of a "trial and error" procedure that must be carried out for every polymer.

Despite the fact that the influence of each variable on fibre morphology is often ambiguous, a number of literature studies have identified few general rules that can be reasonably applied to all ES systems. It is well-established that solution parameters, which rely on polymer and solvent properties, are the most important ones in affecting fibre morphology. Firstly, polymer molecular weight and solution concentration define polymer processability into fibrous meshes. Indeed, these two parameters determine the presence of polymeric chain entanglements that provide a viscoelastic polymer network, essential to prepare ES fibres. For a given polymer, a progressive change from individual electrosprayed beads to continuous electrospun fibres is usually observed as polymer concentration increases (see Fig. 3.12 as an example). It turns out that, for a specific polymer solution, a minimum concentration is required to obtain fibres because, below this value, viscoleastic forces are not high enough to allow the generation of continuous fibres and polymer is collected in the form of beads. Some mathematical approaches have been developed for relating the concentration regime and electrospinnability of polymer solutions [40, 41].

In addition to polymer concentration and molecular weight, the selected solvent system profoundly influences fibre morphology [42, 43]. Indeed, electrical properties and surface tension of the solution interplay in controlling formation and evolution of the ES jet and they are key factors in determining fibre morphology.

ES is considered a very promising scaffold fabrication technology mainly because it produces topological biomimetic scaffolds. However, the great interest in this technology also arises from several other advantages not displayed by other scaffold fabrication techniques:

- Practically all kind of polymers with high enough molecular weight can be electrospun into fibres.
- Tuning process parameters enables to regulate fibre morphology in terms of fibre dimension, presence of beads and fibre surface porosity.
- Fibre deposition and mutual fibre orientation can be easily controlled by acting on the collector.
- Blended polymer fibres, drug-loaded and particle-loaded fibres can be easily produced. It is also possible to concomitantly electrospin different fibres to obtain "composite" non-woven mats.

Fig. 3.12 Representative SEM images of P(D,L)LA in DMF that illustrate fibre morphology changes as a function of polymer concentration: **a** at low concentration beads deposit on the collector, **b** bead-on string fibres are obtained when concentration is increased and **c** at higher concentration bead-free fibres generate

Given the several advantages presented by this technique, the present research was mainly focused on the implementation of an ES apparatus for the production of many different ES scaffolds with highly reproducible morphology and on their application in TE.

A process parameter optimization study is presented in the following section as an example, whereas next sections describe the main features of the ES scaffolds produced in the course of the present research project.

3.2.1 Process Conditions Optimization: The Example of P(PDL-CL)

According to the above discussed considerations, the approach to optimize ES parameters was always a "trial-and-error" procedure applied to every polymer as follows. First, given the importance of solution parameters, different polymer concentrations and several solvent systems were investigated to address their influence on fibre morphology in terms of viscoelastic and electrical properties and of solution surface tension. Once a proper polymer solution was selected, instrumental parameters were then optimized in order to obtain bead-free submicrometer fibres. The optimization procedure of solution and instrumental parameters for ES P(PDL-CL) copolymer is discussed as an example (environmental parameters were maintained constant, T = 25 ± 2 °C, RH = 40 ± 3%).

Table 3.2 lists physical properties [44] of the most common solvents employed for ES aliphatic polyesters and their ability to dissolve the copolymer P(PDL-CL).

Among the different solvents tested, CLF is the only pure organic solvent that dissolves the copolymer at RT. Therefore, solutions of P(PDL-CL), were initially prepared in CLF. Figure 3.13 shows SEM images of P(PDL-CL) electrospun from CLF solutions with increasing polymer concentration. At the lowest polymer concentration (6% w/V, Fig. 3.13a), the polymer collected was in the form of beaded nanofibres. This type of defect is often eliminated by increasing polymer concentration which is concomitantly accompanied by increasing of fibre diameter [40, 45]. Indeed, Fig. 3.13 shows that, by increasing the solution concentration from 6 to 10% w/V, fibres got thicker, although a high density of elongated beads still remained (Fig. 3.13b). Additional experiments carried out using a 10% w/V

Table 3.2 Physical properties of selected solvents [44] and their ability to solubilized P(PDL-CL)

Solvent	PPDL solubility test[a]	Boiling point (°C)	Dielectric constant (dyn/cm)	Surface tension (mN/m^2)
Chloroform (CLF)	Soluble	61	4.8	26.6
Dichloromethane (DCM)	Insoluble	40	8.9	27.2
1,1,3,3,3-Hexafluoro-2-propanol (HFIP)	Insoluble	58	16.7	16.1
2-Chloroethanol	Insoluble	129	25.0	38.9
N,N-dimethylformamide (DMF)	Insoluble	153	38.3	35.7
Acetone	Insoluble	56	21.0	22.7
Tetrahydrofuran (THF)	Insoluble	65	7.5	26.4
Methanol (MetOH)	Insoluble	65	33.0	22.1

[a] soluble when 3% (w/V) solution is optically clear at room temperature, insoluble when 3% (w/V) solution is not optically clear at room temperature

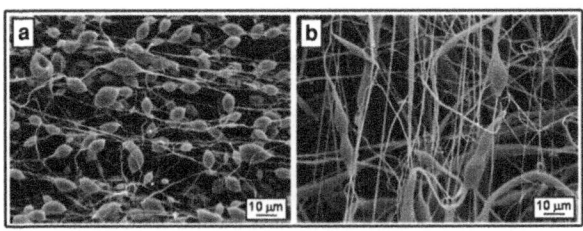

Fig. 3.13 SEM micrographs of P(PDL-CL) electrospun from CLF solutions at different polymer concentrations: **a** 6% w/V and **b** 10% w/V (ΔV = 16 kV, R = 0.01 ml/min, d = 15 cm)

P(PDL-CL) solution in CLF, by changing instrumental parameters (i.e. applied voltage, needle to collector distance and solution flow rate), did not significantly improve fibre morphology. Fong et al. [46] hypothesized that beads generate when surface tension is intermittently higher than electrostatic force and authors stated that balance of the two opposite effects is predominantly determined by viscosity, charge density and solution surface tension. Specifically, high viscosity, high charge density and low surface tension lead to defect-free fibres. In the case of P(PDL-CL)/CLF solutions, Fig. 3.13 shows that the increase of solution viscosity did not prevent bead formation. Therefore, alternative mixed solvent systems were investigated in order to modify electrical properties and surface tension of ES solutions.

Among the solvents reported in Table 3.2, DMF exhibits the higher dielectric constant. A small amount of it added to a P(PDL-CL)/CLF solution is expected to increase the overall dielectric constant of the polymer solution, so that the latter can carry a higher amount of charge [43] (assuming that solvent properties do not change dramatically upon addition of P(PDL-CL) [47]). However, even a small amount of DMF added to a P(PDL-CL)/CLF solution resulted in polymer precipitation. Hence, the CLF/DMF mixed solvent was discarded.

HFIP possesses surface tension and dielectric constant values that might be a good compromise between the demand to increase the charge density and to decrease surface tension of the polymer solution in order to eliminate beads. Indeed, a jet carrying many charges should act against the tendency of the surface tension to maintain the bead shape and it is expected to easily elongate. Another interesting solvent is CE that, despite having a higher surface tension than HFIP, possesses a higher dielectric constant. Figure 3.14 shows fibres obtained from 6% w/V P(PDL-CL) in mixed solvent of CLF/HFIP (80/20 by volume, Fig. 3.14b) and CLF/CE (80/20 by volume, Fig. 3.14c), together with P(PDL-CL) obtained from 6% w/V in CLF for comparison (Fig. 3.14a) (keeping constant all the instrumental parameters). It is evident that the addition of a co-solvent possessing a high dielectric constant helps in obtaining less beaded fibres. However, CE has a better effect compared to HFIP in reducing bead density, probably thanks to its higher dielectric constant.

Fig. 3.14 SEM micrographs of P(PDL-CL) electrospun from 6% w/V in: **a** CLF, **b** CLF/HFIP = 80/20 and **c** CLF/CE = 80/20 (ΔV = 16 kV, R = 0.02 ml/min, d = 15 cm)

The influence of electrical properties on fibre morphology is even more evident when the amount of CE is increased to the detriment of CLF. Figure 3.15 shows SEM images of fibres obtained from 6% w/V P(PDL-CL)/CLF solutions containing different amounts of CE. By increasing CE content in the mixed solvent from 20 to 30% and 40% (Fig. 3.15a–c, respectively), the obtained fibres get thinner.

A further decrease of fibre dimension can be achieved by decreasing polymer concentration. Figure 3.16 shows P(PDL-CL) fibres obtained by changing the

Fig. 3.15 SEM micrographs of P(PDL-CL) electrospun from 6% w/V in: **a** CLF/CE = 80/20, **b** CLF/CE = 70/30 and **c** CLF/CE = 60/40 (ΔV = 13 kV, R = 0.01 ml/min, d = 15 cm)

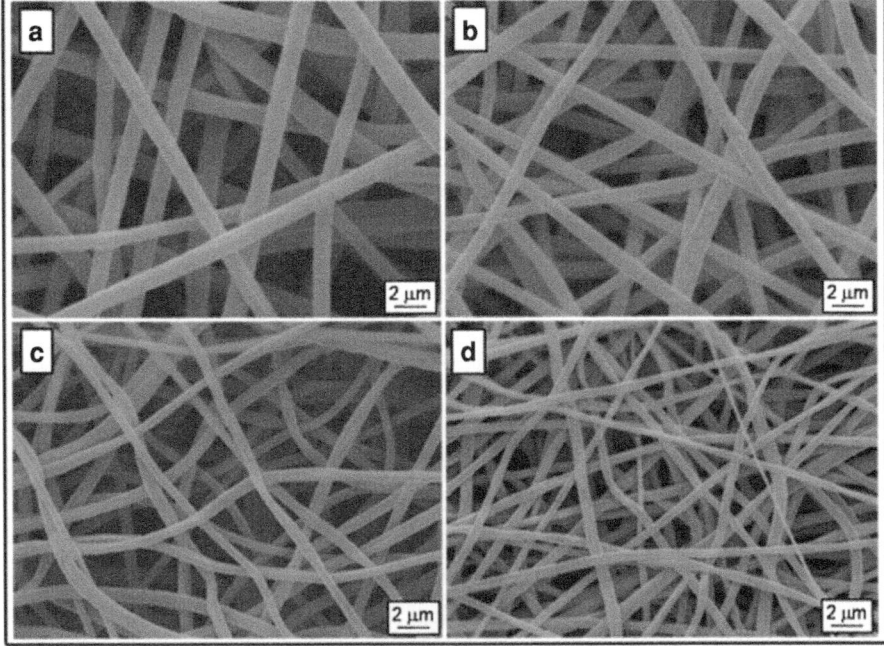

Fig. 3.16 SEM micrographs of P(PDL-CL) electrospun from a mixed solution of CLF/CE = 70/30 (by volume) at: **a** 8% w/V, **b** 7% w/V, **c** 6% wt/V and **d** 5% w/V (ΔV = 13 kV, R = 0.01 ml/min, d = 15 cm)

polymer concentration in CLF/CE (70/30 by volume) from 8% w/V down to 7%, 6% and 5% w/V. The obtained fibres had a diameter distribution of 1080 ± 260 nm, 820 ± 130 nm, 730 ± 180 nm and 500 ± 130 nm respectively.

Although solution parameters are the most important ones in determining polymer electrospinnability and fibre morphology, the latter is also affected by instrumental parameters, even if to a limited extend. Figure 3.17 shows P(PDL-CL) fibres obtained from the same polymer solution but changing the solution flow rate. The flow rate increase from 0.01 to 0.02 ml/min led to a slight increase of fibre dimension from 500 ± 130 to 570 ± 170 nm. Commonly, high flow rate enables the generation of thicker fibres, since it provides a larger amount of solution to be stretched [45].

Figure 3.18 shows the effect of needle-to-collector distance on P(PDL-CL) fibre morphology: by increasing needle-to collector distance from 15 to 20 cm, fibres got thicker (diameter distribution changed from 500 ± 130 to 630 ± 180 nm). Needle-to collector distance influences both the time of jet elongation and the electrical field, therefore the jet elongation force. Its effect can be reasonably

Fig. 3.17 SEM micrographs of P(PDL-CL) electrospun from a 5% w/V mixed solution of CLF/CE = 70/30 at: **a** R = 0.01 ml/min and **b** R = 0.02 ml/min (Δ = 13 kV, d = 15 cm)

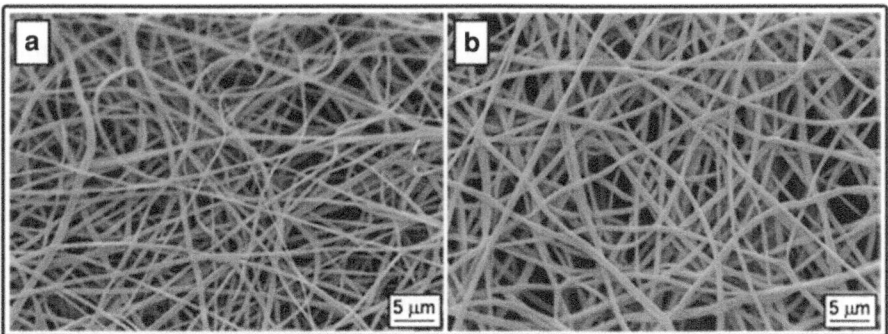

Fig. 3.18 SEM micrographs of P(PDL-CL) electrospun from a 5% w/V mixed solution of CLF/CE = 70/30 at: **a** d = 15 cm and **b** 20 cm (ΔV = 13 kV, R = 0.01 ml/min)

explained by taking into account the two extreme situations: very low distances generate high electrical fields that stretch the jet for very short times whereas very high distances generate low electrical fields to stretch fibres over long distances. In both cases thicker fibres are expected than those obtained at intermediate needle-to-collector distances where strong enough fields for long enough time are applied to stretch the fibres. Obviously, the best range of needle-to collector distance strictly depends on all other ES conditions, in particular on the applied voltage.

Experiments carried out on 5% w/V P(PDL-CL) solution of in CLF/CE = 70/30 by changing the applied voltage did not provide significantly different fibre dimensions, so that the applied voltage seemed to have no evident effect on fibre morphology. Again, it is pointed out that each ES variable can affect fibre morphology depending on the set of conditions used. It can be concluded that the applied voltage does not have any effect within the range of ES conditions investigated. However, in general, applied voltage can have a deep influence on fibre morphology, especially on the occurrence of bead-on-string fibres. Indeed, it directly controls the amount of charge on solution drop that, overcoming the surface tension, is responsible to increase the jet surface area [48].

Optimization study of process parameters applied to P(PDL-CL) copolymer led to identify the following ES conditions for the production of defect-free sub-micrometric fibres:

- P(PDL-CL) solution: 5% w/V in CLF/CE (70/30 by volume),
- applied voltage: $\Delta V = 16$ kV,
- solution flow rate: R = 0.01 ml/min,
- needle to collector distance: d = 15 cm.

3.2.2 Electrospun Polymers

ES scaffolds from different degradable polyesters are presented in this section. In the course of the present research both commercial and non-commercial polyesters were electrospun into fibres. The commercial polymers used in the course of this Ph.D. (i.e. polylactides and their copolymers and poly(ε-caprolactone)) have been more intensively investigated in the literature as bioresorbable materials than any other degradable polymer. These materials have been already successfully used in approved medical devices. Indeed, they are biocompatible and non-toxic and their use is approved in most of developed countries. Their intensive use in biomedical research is justified by the fact that bringing to market implantable scaffolds prepared from approved polymers is quicker and cheaper than using novel polymers whose biocompatibility is still not attested.

ES conditions parameters were optimized in order to obtain defect-free fibres by evaluating fibre morphology by Scanning Electron Microscopy. The optimization study was carried out for every polymer as previously described for P(PDL-CL). Both pristine polymers and the corresponding ES fibres were characterized by

Thermogravimetric Analysis (TGA) and Differential Scanning Calorimetry (DSC). TGA did not reveal residual solvents in the ES fibres described below and confirmed that ES process did not affect polymer thermal degradation behaviour. DSC was used to evaluate the ratio of crystal to amorphous phase in ES fibres. From DSC first scans emerged that, in some case, this ratio differed from the one of the pristine material. On the contrary, DSC second scans of pristine polymers and of the corresponding fibres were always identical.

3.2.2.1 Polylactide

Lactic acid is in many natural products. It is an optically active molecule that exists both as D or L stereoisomer. Commonly, lactide polymers are synthesized via ring opening polymerization of the cyclic dimer lactide (either (L,L) or (D,L)-lactide) and polylactide properties vary depending on the ratio of the two stereoisomers. The optically pure polymers are semicrystalline and, if the other enantiomer molar content is less than 20%, macromolecules are still able to crystallize [49].

Since polylactides are relatively unstable in the presence of moisture, their use was not widespread till 1960, when their benefits in the biomedical field became evident. Polylactides are biodegradable polymers that offer the great advantage of maintaining their functionalities for a limited period of time, after which they are naturally resorbed by the body. Semicrystalline polylactides are mainly used as load-bearing implants in orthopaedic fixations (e.g. screws, pins) [50] whereas amorphous polylactides are often employed in drug delivery devices [51] where a homogeneous dispersion of the active species within a monophasic matrix is required. With the development of TE principles, polylactide and, more in general, lactide-based copolymers became perhaps the most commonly used polymers for scaffold fabrication. Moreover, polylactides are employed for packaging, agricultural and disposable applications. The intrinsic biodegradability and the derivation from renewable resources allow, on one hand, to reduce the social waste problem and, on the other hand, to reduce the use of carbon fossil-derived materials.

Amorphous poly(D,L)lactide and semicrystalline poly(L)lactide were both electrospun by using process conditions listed in Fig. 3.19. The figure also shows SEM micrographs and fibre diameter distributions of the obtained scaffolds.

Table 3.3 provides calorimetric data obtained from DSC first scans of P(D,L)LA and P(L)LA samples. Given the absence of backbone stereoregularity, P(D,L)LA is not expected to develop a crystalline phase during ES process. On the contrary, P(L)LA may be able to crystallize. P(L)LA starting sample is semicrystalline, with a melting endotherm peak at $T_m = 162\ °C$ and a melting enthalpy $\Delta H_m = 35\ J/g$, corresponding to a crystallinity degree $\chi_c = 37\%$ (calculated according to Eq. 2.2, assuming $\Delta H_m° = 93\ J/g$ [52]). DSC first scan of P(L)LA fibres shows a cold crystallization exotherm peak followed by a melting endotherm of the same entity ($\Delta H_c = \Delta H_m = 31\ J/g$). Therefore, P(L)LA fibres reported in Fig. 3.19 appears to

Fig. 3.19 P(D,L)LA and P(L)LA ES scaffolds: process conditions, SEM micrographs and fibre diameter distributions

Table 3.3 Calorimetric data of P(D,L)LA and P(L)LA samples

Polymer	T_g (°C)	ΔC_p (Jg^{-1}°C^{-1})	T_c (°C)	ΔH_c (Jg^{-1})	T_m (°C)	ΔH_m (Jg^{-1})	$\Delta H_m - \Delta H_c$ (Jg^{-1})	χ_c^d
P(D,L)LA as received[a]	54	0.54	–	–	–	–	–	–
ES P(D,L)LA[a, c]	54	0.39	–	–	–	–	–	–
P(L)LA as received[b]	64	0.15	–	–	162	35	35	37
ES P(L)LA[b, c]	61	0.49	86	31	159	31	0	0

[a] First scan from −50 to 120 °C, heating at 20 °C/min
[b] First scan from −50 to 200 °C, heating at 20 °C/min
[c] ES mats obtained by using process conditions listed in Fig. 3.19
[d] Calculated according to Eq. 2.2, by assuming $\Delta H_m° = 93$ J/g [52]

be completely amorphous. P(L)LA is a slow crystallisable polymer compared to many conventional thermoplastic polymers [53, 54]. During ES process, polymer chains have little time to organize in a crystal structure before the occurring of fibre solidification and polymer crystallization can be inhibited if crystallization rate is low.

3.2.2.2 Poly(lactide-*co*-glycolide) Copolymers

Properties of polylactide can be adapted to a wide range of possible applications through copolymerization of lactide with other monomers, the most intensively investigated one being glycolide. PLAGA copolymers are biomedical commercial products since 1970 and they have been developed mainly for resorbable sutures (trade names Vicryl and Polyglactin 910) but they are also employed in craniomaxillofacial reconstruction and other orthopaedic applications [50]. The lactide-glycolide molar ratio regulates mechanical properties and phase morphology but, above all, it affects the rate of degradation. The presence of the hydrophilic GA units increases the water uptake of the copolymers, thus accelerating ester cleavage and giving the possibility to tune resorption rate of the implant in the organism. Amorphous copolymers with high glycolide content are commonly resorbed within 1–2 months. When glycolide content is low, resorbtion times increase up to 6 months, a time still considerably lower than that required for resorption of the homopolymer polylactide (more than 1 year) [51]. However, the intensive clinical use of PLA and PLAGA copolymers highlighted that inflammatory reactions are the most serious complications associated with these materials. Inflammations are probably related to the release of acidic products as a consequence of polymer degradation. This problem might be reduced by using more hydrophobic polymers that degrade slowly, thus obtaining the release of a lower amount of acid products.

The copolymers employed in the present research have different lactide-glycolide molar ratio: $PLA_{90}GA_{10}$, $PLA_{75}GA_{25}$, $PLA_{65}GA_{35}$ and $PLA_{50}GA_{50}$. They were all synthesized from (D,L)-lactide apart from the copolymer $PLA_{90}GA_{10}$ which was synthesized from the stereoregular (L,L)-lactide. Therefore, $PLA_{75}GA_{25}$, $PLA_{65}GA_{35}$ and $PLA_{50}GA_{50}$ copolymers re not expected to develop a crystal phase whereas $PLA_{90}GA_{10}$ may be able to develop a crystal phase given the stereoregularity of the lactide units. With the increase of the amount of glycolide unit, that has higher mobility then lactide unit, the copolymer T_g decreases as reported in Fig. 3.20 (T_g of polyglycolide taken from [51]).

Figure 3.21 reports optimized process conditions employed in the present research to electrospin $PLA_{90}GA_{10}$, $PLA_{75}GA_{25}$, $PLA_{65}GA_{35}$ and $PLA_{50}GA_{50}$.

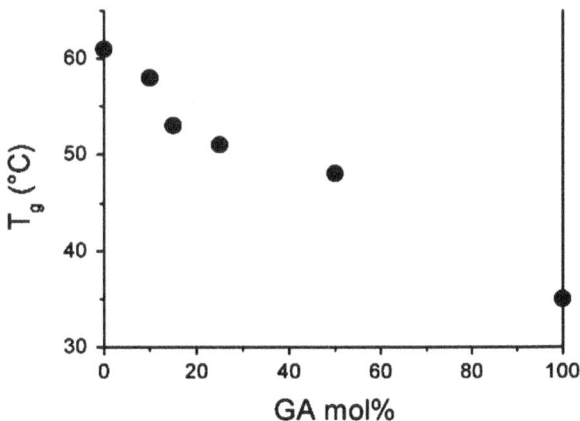

Fig. 3.20 Glass transition temperature of PLAGA copolymers as a function of glycolide content

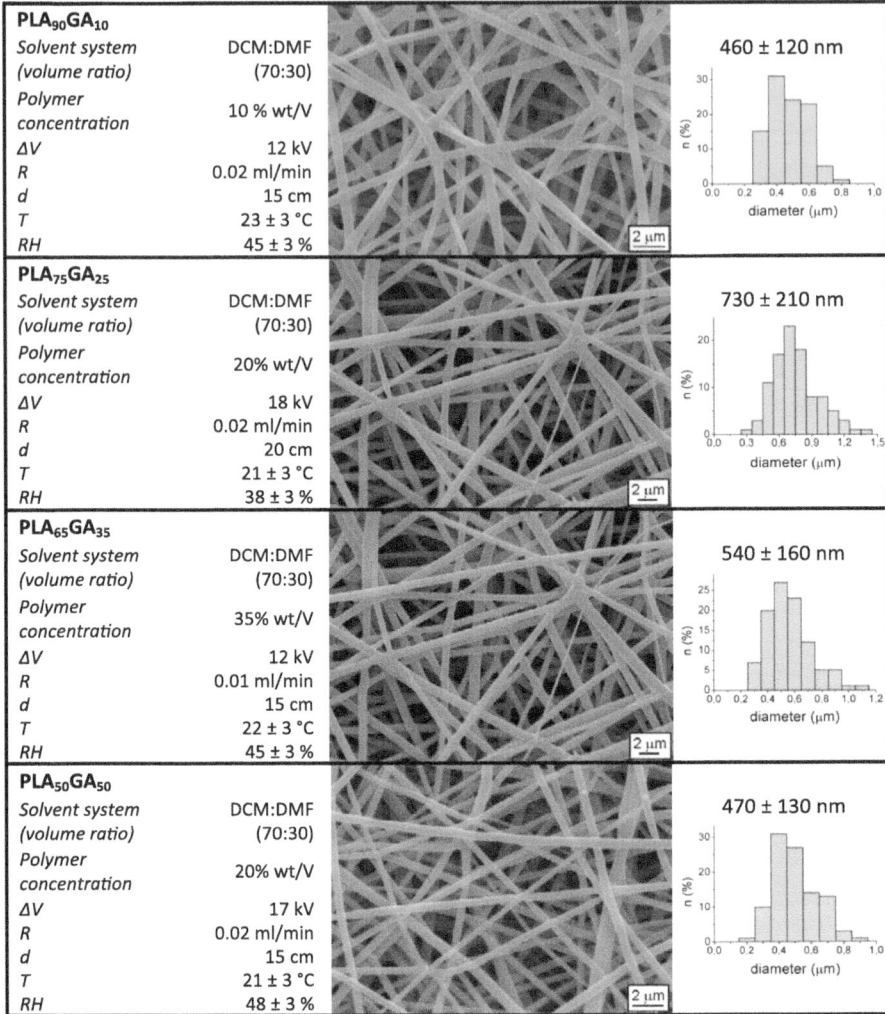

Fig. 3.21 PLAGA ES scaffolds: process conditions, SEM micrographs and fibre diameter distributions

DSC analysis showed that all ES fibres shown in Fig. 3.21 are completely amorphous.

3.2.2.3 Poly(lactide-*co*-trimethylene carbonate)

Another interesting copolymer of lactic acid is P(LA-TMC). The addition of TMC units decreases T_g and confers elastic properties to the materials making P(LA-TMC) copolymers good candidates for soft tissue reconstruction [55, 56].

P(LA-TMC) employed in the present research is a random copolymer of (L,L)-lactide with a TMC content of 30% in weight. It has a T_g around 34 °C, close to the physiological temperature. Figure 3.22 shows sub-micrometric P(LA-TMC) fibres obtained by using optimized process conditions [57].

P(LA-TMC) ES fibres are completely amorphous with a T_g around 34 °C (by DSC first scan, heating at 20 °C/min, not shown). Shape memory properties of P(PLA-TMC) ES mats have been recently observed in a preliminary study carried out in the laboratory [57]. Since the switching temperature associated with the shape recover is closed to physiological conditions, these ES scaffolds are potentially extremely interesting.

3.2.2.4 Poly(ε-caprolactone)

Poly(ε-caprolactone) is a biocompatible polymer more hydrophobic than both PLA and PLAGA copolymers, it degrades slower and it is therefore suitable for long-term applications. In particular a long-term implant called Capronor has been intensively studied as contraceptive deliverer [58].

Figure 3.23 reports SEM micrographs of micrometric PCL fibres.

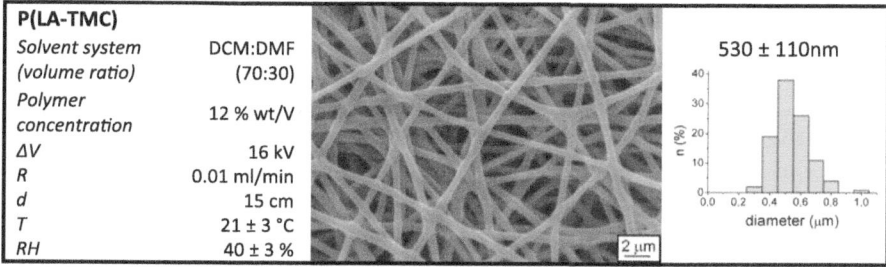

Fig. 3.22 P(LA-TMC) ES scaffolds: process conditions, SEM micrograph and fibre diameter distribution

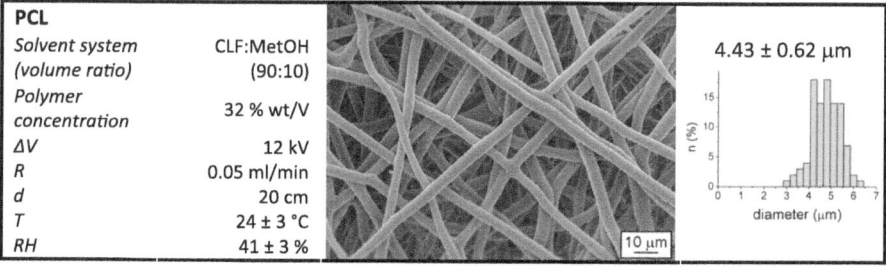

Fig. 3.23 PCL ES scaffolds: process conditions, SEM micrograph and fibre diameter distribution

Table 3.4 Calorimetric data of PCL samples

Polymer	T_g (°C)	ΔC_p (Jg^{-1}°C^{-1})	T_c (°C)	ΔH_c (Jg^{-1})	T_m (°C)	ΔH_m (Jg^{-1})	$\Delta H_m - \Delta H_c$ (Jg^{-1})	χ_c^c
PCL as received[a]	−59	0.08	–	–	62	88	–	63
ES PCL[a, b]	−63	0.10	–	–	57	77	–	55

[a] First scan from −100 to 130 °C, heating at 20 °C/min
[b] ES mats obtained by using process conditions listed in Fig. 3.23
[c] Calculated according to Eq. 2.2, by assuming $\Delta H_m° = 140$ J/g [61]

PCL is a semicrystalline polymer that, unlike P(L)LA, is able to crystallize during ES process. Table 3.4 compares calorimetric data obtained from first DSC scans of the pristine sample and of ES PCL fibres. The crystallinity degree of ES fibres appears to be lower than that of the "as received" sample. This result is not surprising since the fast ES process is expected to limit crystal formation [59, 60] or even to totally inhibit crystallization, as above reported in the case of P(L)LA.

3.2.2.5 Poly(ω-pentadecalactone) and Poly(ω-pentadecalactone-*co*-p-dioxanone)

PPDL and its copolymers are degradable materials that, despite not being medically approved yet, display different and extremely interesting properties with respect to the commonly used approved biomaterials. Firstly, given the presence of ω-pentadecalactone units with 14 methylene units per ester group, they are slowly degradable polymers. This might reduce inflammatory reactions caused by fast release of acidic by-products and they might be particularly interesting for long-term applications. Moreover, PDL has been successfully copolymerized with comonomers such as trimethylene carbonate [62], ε-caprolactone [16], and *p*-dioxanone [63], thus enabling material scientists to tailor corresponding hydrolysis rate and biomaterial physical properties for targeted applications.

In the present research PPDL and a P(PDL-DO) copolymer have been electrospun, as well as a P(PDL-CL) copolymer already described in the previous Sect. 3.2.1. Both PPDL homopolymer and its copolymers used in the present research are highly crystalline materials. The peculiar crystallization of P(PDL-CL), where PDL and CL units cocrystallize in the same lattice [16] displaying a isomorphic behaviour, has been already discussed in Sect. 3.1. Similarly to P(DL-CL), in P(DL-DO) the two monomeric units cocrystallize in the same lattice, with a behaviour, in this case, typical of a isodimorphic system [63]. Figure 3.24 lists optimized process conditions for obtaining sub-micrometric bead-free PPDL and P(PDL-DO) fibres together with corresponding SEM micrographs and fibre diameter distributions. Calorimetric analysis of fibres reported in Fig. 3.24 shows that both PPDL and P(DL-DO) copolymer are able to crystallize during fibre solidification. Similarly to PCL, PPDL and P(DL-DO) fibres are less crystalline than the corresponding starting polymers (by DSC analysis).

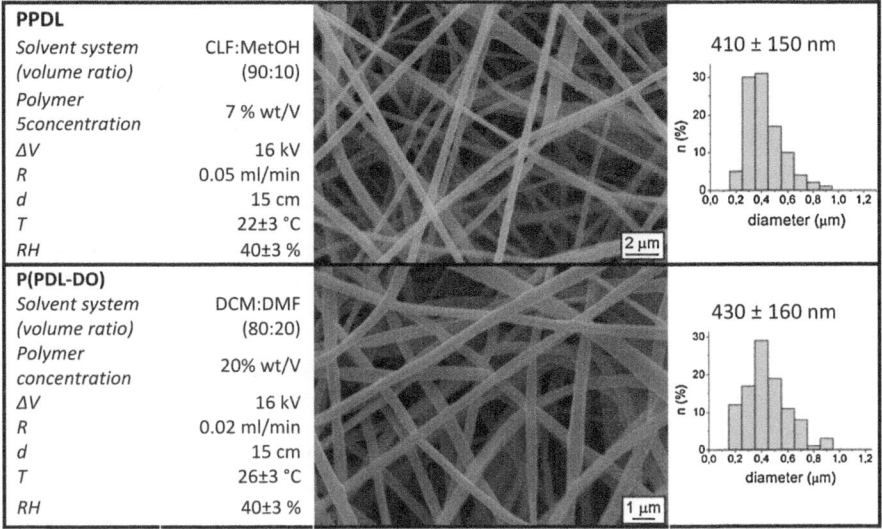

Fig. 3.24 PPDL and P(PDL-DO) ES scaffolds: process conditions, SEM micrographs and fibre diameter distributions

3.2.3 Fibre Morphologies

The adjustment of ES parameters allows obtaining many different fibre morphologies in terms of presence of beads and bead shape, fibre porosity and dimension. Section 3.2.1 has already introduced the so-called "bead-on-strings" morphology whose formation has been attributed to surface tension forces intermittently overcoming electric and the viscoelastic forces during jet elongation [46]. Commonly, beads have either spherical or "spindle-like" aspects (Fig. 3.25a, b, respectively) but also more "exotic" shapes can be obtained, such as hollow beads and porous beads (Fig. 3.25c, d, as examples). Cup-shaped beads and "prune-like" beads have also been reported in the literature [48, 64]. Commonly, beads are considered defects along the fibre and process optimization studies usually aim at reducing their density. Indeed, high bead densities are responsible of worsening mechanical resistance of fibrous mats, sometimes preventing mat detachment from the collector without damaging it. However, the presence of beads along the fibres might be useful for diffusion-controlled drug release application, even if this possibility has not been explored yet. Indeed, a drug located at the bead core should diffuse out from the polymer more slowly than a drug located inside the thin fibre.

As already discussed in the previous sections, the control of ES parameters allows to tailor fibre diameters within a dimensional range from several microns down to few hundreds of nanometers. As a matter of fact, fibre thickness determines pore dimension in the fibre mat: non-woven mats with wider interstices can

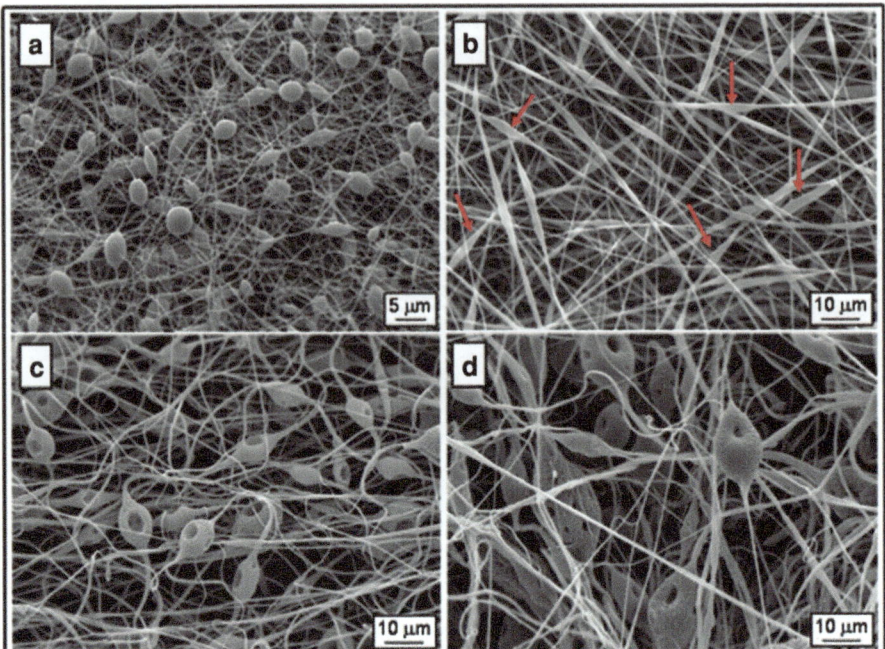

Fig. 3.25 Representative SEM micrographs of fibres displaying: **a** spherical beads, **b** "spindle-like" beads (*red arrows*), **c** hollow beads and **d** porous beads

be fabricated by increasing fibre diameter [1]. In view of using ES materials as cell culture supports, pore dimension is particularly important. Indeed, cells can easily penetrate a 3D structure with pore dimension comparable with that of cells, whereas the infiltration in smaller pores can be limited.

Fibre diameter also modulates fibre surface area: thinner fibres lead to porous mats with higher surface to volume ratio. In biomedical applications, surface area is a key factor since cell-device interactions occur at biomaterial interface. Moreover, surface area is particularly relevant when scaffolds are loaded with drugs that are intended to be released in a controlled manner.

Besides fibre diameter, surface area is also determined by fibre surface porosity and roughness. ES technique allows to produce porous fibres that, as a matter of fact, display a larger surface area with respect to smooth fibres of similar dimension. Pore generation at fibre surface can be achieved by appropriately changing ES conditions, such as solvent system and relative humidity. Figure 3.26a, b show porous fibres obtained by electrospinning a P(L)LA 20% w/V in DCM and in CLF respectively. DCM generates thinner fibres than CLF because of its higher dielectric constant. Interestingly, depending on the solvent, fibres display different porosity patterns. It has been speculated that pores generate as a consequence of phase separation during fast solvent evaporation [37]. This hypothesis is supported by the fact that the addition of a high boiling point solvent—such as DMF—to both

ES solutions inhibits pore formation (not shown). Fibre porosity can be also controlled by relative humidity: surface pore density can be reduced by decreasing moisture (see Fig. 3.27). This finding is attributed to the condensation of water droplets at the surface of the polymer jet, which is undergoing cooling as a

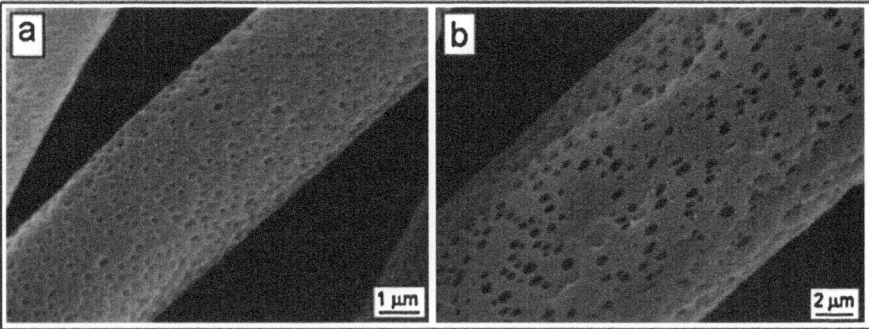

Fig. 3.26 SEM micrographs of porous P(L)LA fibres obtained by electrospinning a solution of P(L)LA 20% w/V in: **a** DCM and **b** CLF

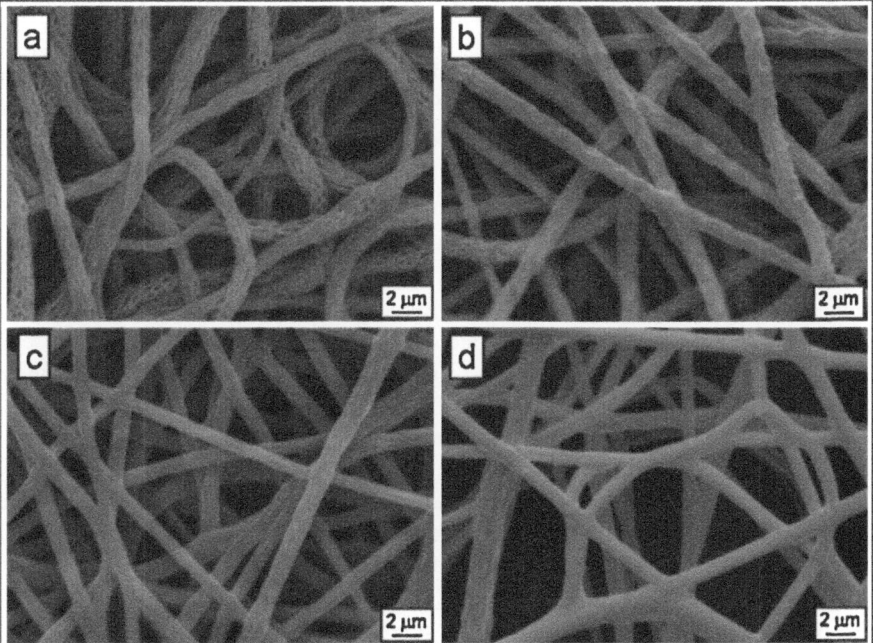

Fig. 3.27 SEM micrographs of P(PDL-DO) fibres obtained by electrospinning a solution of P(PDL-DO) 30% w/V in DCM/DMSO (80/20 by volume) at: **a** RH = 55%, **b** RH = 45%, **c** RH = 35% and **d** RH = 25%. Relative humidity was the only parameter varied during this experiment

consequence of fast solvent evaporation. The imprints of these droplets, whose density is higher at high RH values, leave behind a porous surface [37, 65].

3.2.4 Fibre Deposition Geometries

In the most common ES setup fibres are collected on a stationary plane metallic target as non-woven meshes of randomly oriented fibres. The poor control of spatial deposition distribution is a direct consequence of the chaotic path of the charged jet that travels towards the metallic plate. Control of the jet path and of fibre alignment has been faced by several researchers and it has been summarized in a comprehensive review by Teo et al. [66], who highlighted the key role of material and geometry collector in guiding fibre deposition. Fibre alignment is commonly achieved by using either a high speed rotating tool (e.g. a cylinder [67] or a sharp edged disk [68]) or auxiliary guiding electrodes [69–71]. Then special collector geometries have been used to achieve fibre alignment, such as parallel conducting strips spaced apart by an insulating gap [72, 73] or multiple gold electrodes deposited on quartz wafers [72, 74]. However, these latter approaches enable to align fibres only in a confined limited space. Fibre patterning has been achieved by using conducting grids [75, 76], although the obtainment of complex patterns still remains a challenge.

The possibility to produce ES scaffolds having controlled fibre deposition geometries opens new routes towards the investigation of cell-material interactions and of cell behaviour addressed by substrate morphology.

3.2.4.1 Aligned Fibres

In the course of the present research, highly aligned fibres were produced by using a metallic cylindrical rotating collector specifically designed for this scope (Fig. 3.28). A thermoplastic solvent-resistant (POM-H®) support that can locate cylindrical targets of different dimensions (maximum length = 22 cm, maximum radius = 4 cm) is used. The engine enables cylinder rotation up to a maximum angular speed ω = 6200 rpm.

It is pointed out that rotational speed can be expressed either as angular speed (ω) or as surface linear speed of the collector (v). By keeping constant the angular speed, the surface linear speed increases with increasing of cylinder radius (Fig. 3.29).

The relation between angular and linear speed implies that, by keeping constant the angular speed (ω), different linear rates (v) can be produced by simply changing the cylinder radius (see Fig. 3.30). In this study higher alignment was achieved by using a larger collector, confirming that surface linear rate of the cylinder is the parameter that controls the degree of fibre alignment.

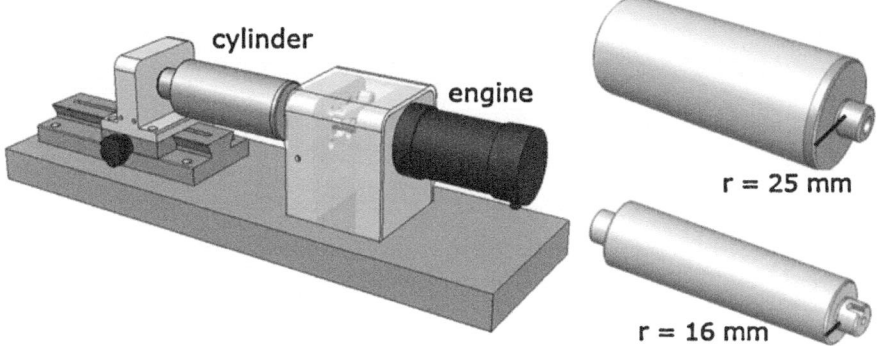

Fig. 3.28 Rotating tool for the production of aligned fibres. Cylinders with different radius can be used

Fig. 3.29 Relation between angular speed (ω) and linear speed (v) of the cylinder. Fibres align parallel to the linear rate, which is tangential to the collector surface

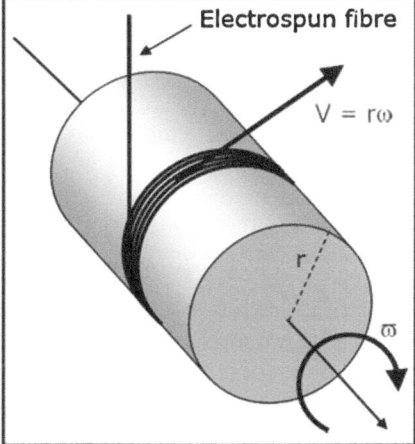

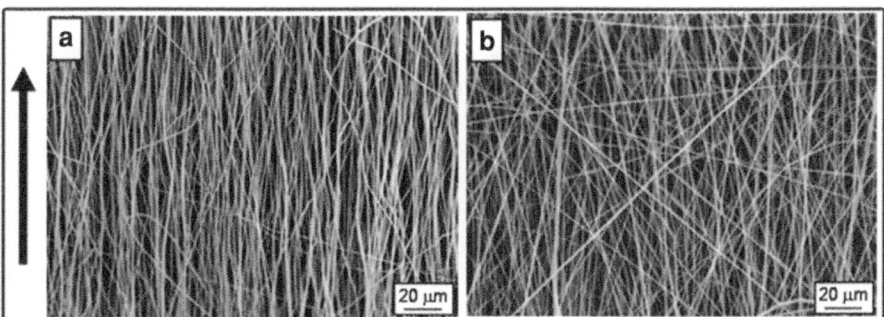

Fig. 3.30 SEM micrograph of P(L)LA fibres deposited on mandrel collectors rotating at angular rate $\omega = 6200$ rpm. **a** mandrel radius = 25 mm, linear rate = 16.2 m/s and **b** mandrel radius = 16 mm, linear rate = 10.4 m/s. The *arrow* indicates the linear rate direction

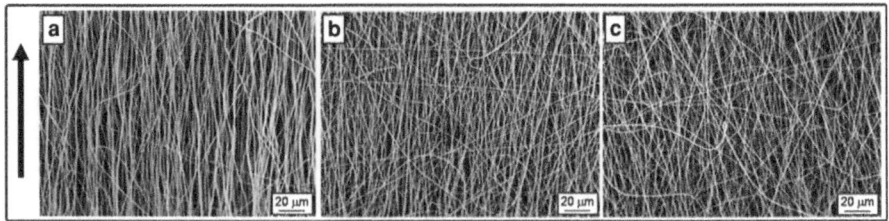

Fig. 3.31 SEM micrograph of P(L)LA fibres deposited on a cylinder collector (r = 25 mm) rotating at linear rate: **a** v = 16.2 m/s, **b** v = 13.9 m/s and **c** v = 9.9 m/s. The *arrow* indicates the linear rate direction

The movement of the cylinder is responsible of mechanical aligning the fibres along the rotation direction. Indeed, during fibre formation, the jet travels at a very high speed and tends to deposit randomly on the collector surface. However, if the collector moves at a speed higher than that of fibre deposition, fibres can wind around the cylinder and align along its circumference [77]. Figure 3.31 shows that fibres, deposited on the same target rotating at different speeds, display a higher degree of alignment when higher rotational speed is used.

3.2.4.2 Patterned Non-woven Mats

Patterned mats can be produced by acting on material collector and geometry. In the course of the present research, non-woven scaffolds, displaying patterns on a different dimensional range, were produced by using two types of target:

- plane collectors made of a conducting metallic substrate coated with a doped dielectric material, supplied by Smaltiflex SpA (Modena, Italy). Fibres deposited on these collectors generate "star-shaped" patterns, where "stars" of about 50 μm in dimension are spaced out hundreds of microns [78];
- plane collectors made of an array of pins fixed on an insulating material. In this case, a macroscopic pattern, where local fibre orientation is controlled by the pin array geometry, is produced.

Collectors supplied by Smaltiflex were made of a metallic substrate coated with an enamel layer[1] [78]. The main enamel components are refractories (quartz, feldspar, clay) and fluxes (borax, soda ash, cryolite, fluorspar), together with opacifiers, colours, floating agents and electrolytes [79, 80]. The interest in coupling ES technology with vitreous enamel is related to the specific electrical properties of enamel. Indeed, enamel is a dielectric material possessing high electrical resistivity that can be modified by varying the content of the alkaline oxides, being alkali metal cations the actual charge carriers.

---

[1] Enamel raw materials were prepared by using a home-made frit with the following composition (in wt%): SiO_2: 43, Al_2O_3:1, B_2O_3: 13, Na_2O: 8, K_2O: 5, ZnO: 1, TiO_2: 22, P_2O_5: 3, F_2: 4, that gives a final white colour coating.

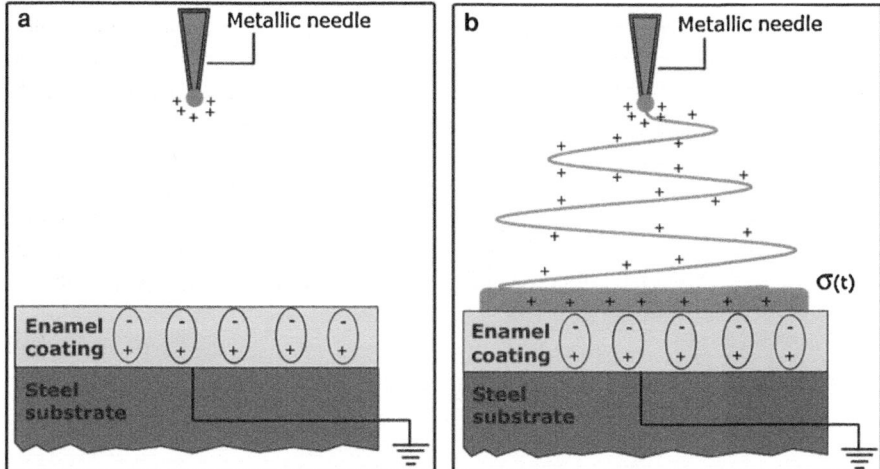

Fig. 3.32 Schematic illustration of the electrostatic phenomena occurring during the ES process using enamelled collectors. **a** starting situation before fibre deposition and **b** after deposition of a layer of polymer fibres (Reprinted from [78], Copyright (2010), with permission from Springer Science)

In general, the use of a dielectric collector affects the fibre discharging process. For the sake of clarity, Fig. 3.32 sketches the electrostatic phenomena occurring when an enamelled steel target is used to collect fibres. When a high voltage difference is applied between the metallic capillary ejecting the polymeric solution and the grounded collector, the enamel dielectric layer is subjected to polarization (Fig. 3.32a). As soon as fibres reach the collector, they tend to rapidly discharge. However, when a uniform layer of fibres is deposited (Fig. 3.32b), fibre discharging becomes less efficient and the collector surface, being covered by partially charged fibres, acquires a positive charge density (σ) that decreases with time due to charge relaxation process across the dielectric coating. The relaxation process is strictly dependent on enamel electrical properties, in particular on its electrical resisitivity.

The positive residual charge density (σ) on collector surface generates a repulsive force towards the incoming fibres, that is larger at higher collector resistivity values. The presence of the residual charge σ mainly affects fibre packing density. Since the electrical resistance of the enamelled collectors increase with enamel coating thickness [78], by using collectors with a chemically identical enamel layer, but with different thickness,[2] it is possible to tailor the fibre packing density. Figure 3.33 shows that mat section displays a less packed fibre density

---

[2] The enamel raw material was prepared by milling the frit for about 11 h in order to obtain a powder that was deposited electrostatically over a 0.8 mm thick steel sheet. Enamel coating thickness was controlled during the deposition process and collectors with different enamel thickness (0.15 and 0.53 mm) were prepared. The coated specimens were fired in a radiant furnace with oxidizing atmosphere at 850 °C for 6.5 min.

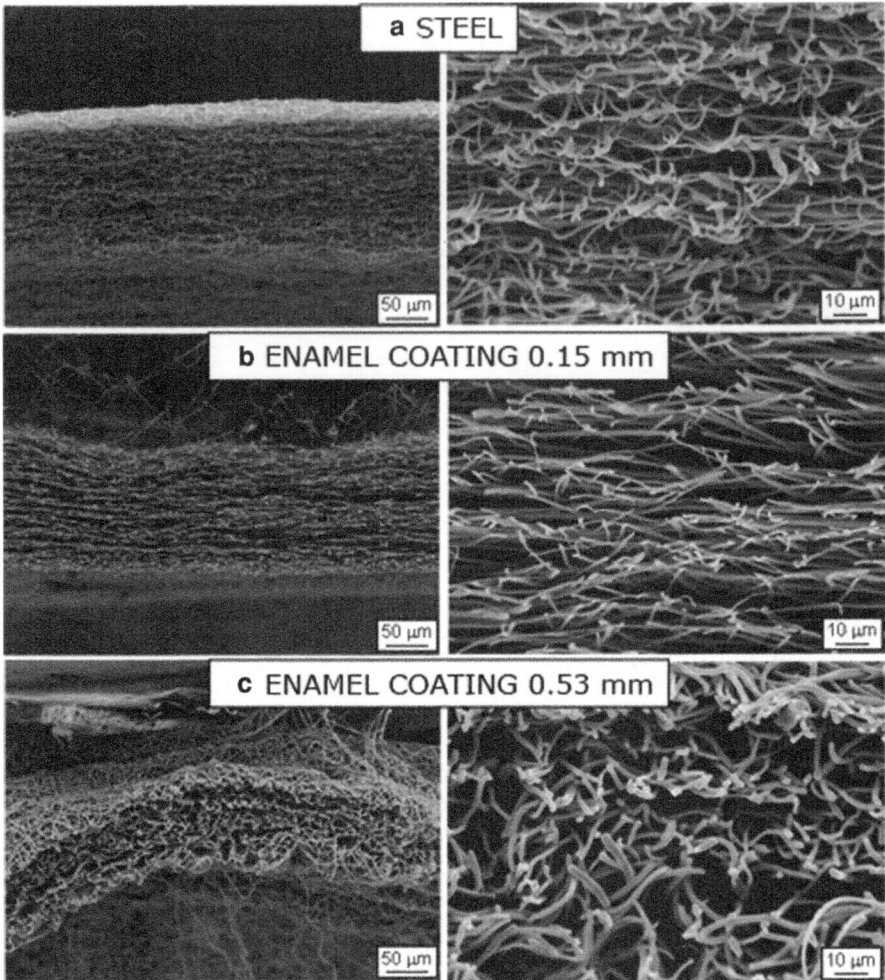

Fig. 3.33 Effect of enamel layer thickness on fibre packing density. SEM micrographs (two different magnifications) of P(L)LA ES mat sections obtained from collectors with different enamel thickness: **a** 0 mm (uncoated steel), **b** 0.15 mm and **c** 0.53 mm (Reprinted from [78], Copyright (2010), with permission from Springer Science)

when the metallic collector is covered with an enamel layer (compare Fig. 3.33a, b) and that fibre packing density is reduced with the increase of dielectric layer thickness (compare Fig. 3.33b, c). This result is due to a slower charge dissipation process occurring when a thicker enamel coating is used, that results in higher repulsive electrostatic forces between the incoming charged fibres and in higher porosity.

The use of the above described enamel material did not induce formation of patterned mats, on the contrary, fibres were collected in the common random configuration. In order to obtain a pattern, the electrical properties of the enamel

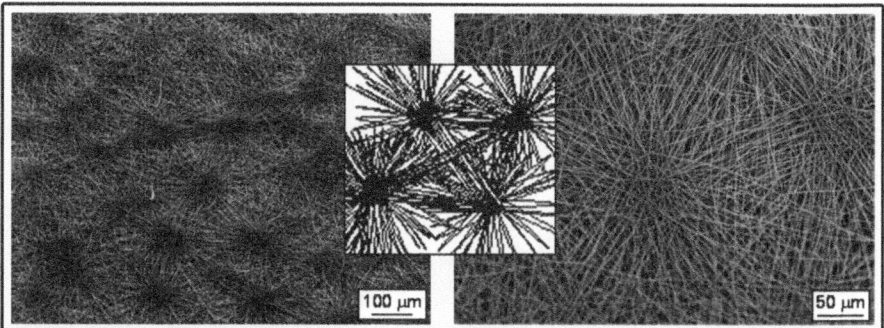

Fig. 3.34 SEM micrographs (two different magnifications) of P(L)LA star-patterned ES mat deposited on enamelled collector containing montmorillonite clay particles (Reprinted from [78], Copyright (2010), with permission from Springer Science)

were locally modified by doping it with proper additives, with the aim of affecting the electrostatic charge diffusion and charge distribution on the collector during the ES process. When montmorillonite clay particles were randomly distributed into the vitreous matrix during collector fabrication,[3] a "star-shaped" pattern was observed on the ES polymer mat deposited (Fig. 3.34). This finding can be associated with the high electrical mobility of the alkaline metals ions present in the montmorillonite, which locally decrease electrical resistivity of enamel. In correspondence to these conducting points the charge density σ is low. Therefore, they act as "potential wells" and aggregation centres for the incoming nanofibres, thus leading to a "star-shaped" deposition.

The second type of collector enabling the obtainment of patterned non-woven mats is composed of: (a) a metallic substrate, (b) regularly spaced metallic pins connected to the grounded metallic substrate and (c) an insulating material laying on the top of the metallic substrate. Figure 3.35 illustrates the target assembly.

Figure 3.36b shows the picture of a non-woven mat detached from the "pin array" target reported in Fig. 3.36a. Pins were positioned on the vertexes of an imaginary square (side length D = 5 mm). In correspondence to the positions of the pins, the mat is thicker because fibres preferentially deposit on the conducting pins. Fibres partially deposit on Teflon sheet, in particular they accumulate both on the sides and on the diagonals of the square, i.e. along lines connecting the conducting pins. Figure 3.36c sketches the peculiar macroscopic pattern of the obtained mat. SEM micrographs of Fig. 3.36 show the mat micro-pattern: fibres tend to spun across the metallic pins, uniaxially aligning along the square sides and

---

[3] Enamel raw material was prepared by milling the frit in water for 4 h and by adding, during the milling process, a montmorillonite clay (Mullenbach & Thewald GMBH, Germany, particle size ca. 50 μm) as doping material. The wet blend containing the suspended clay particles was sprayed over the steel sheet and dehydrated. Enamelled collector with a coating thickness of 0.2 mm was obtained. The collector was fired in a radiant furnace with oxidizing atmosphere at 850 °C for 6.5 min.

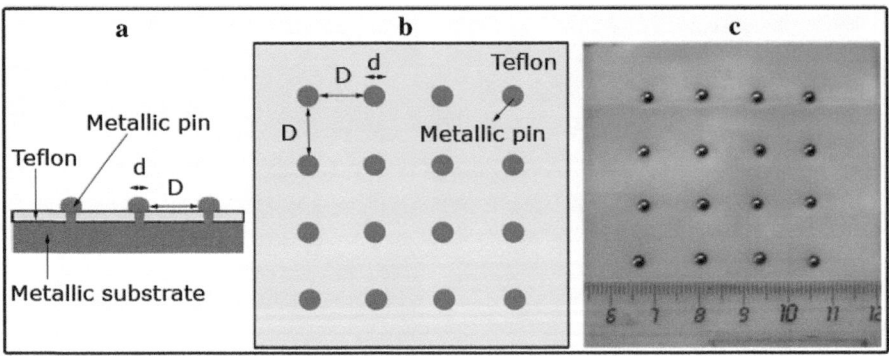

Fig. 3.35 Schematic representation of collectors made of an array of metallic pins (d = 2 mm,) connected to a metallic substrate, placed on a insulating Teflon sheet. **a** side view, **b** front view and **c** picture of a collector where D = 10 mm

generating a cross array in the centre of the square. The collector used allows to locally orient fibres in different directions, depending on the location of the conducting pins. The control of fibre deposition is achieved, in this case, thanks to the action of the electrical field generated between the metallic needle ejecting the polymer solution and the pin array collector. In order to explain its effect, the study by Liu et al. [73] is particularly useful. The authors modelled the electrical field generated by a collector made of two metallic strips separated by an air gap. They found that fibres uniaxially aligned perpendicularly to the strips thanks to the electrical field component parallel to the collector surface that was responsible of stretching the incoming fibres across the conducting strips. Authors pointed out that, when a metallic plate is used, the electrical field does not have a component parallel to the surface collector, thus fibres deposit randomly. Analogously, the peculiar fibre deposition on the "pin array" collectors can be explained according to the electrical field model provided by Liu et al. [73]. Moreover, the pin gap distance D has a remarkable effect on fibre alignment degree. Indeed, by increasing pin separation from 5 mm up to 30 mm, fibre alignment on the sides of the square was enhanced. This result was also found by Liu et al. [73] in the case of metallic strips separated by an air gap. However, the increase of pin distance gradually reduced the fibre deposition on the diagonals of the square, until the cross array fibre morphology in the centre of the square was not generated anymore. Moreover, at even higher pin separation, also fibre deposition across the side of the square is reduced, probably because the electrical field component parallel to the collector surface is too low to stretch the incoming fibres across the conducting pins.

3.2.5 "Composite" Electrospun Scaffolds

Porous scaffolds composed of different types of fibres can be easily produced by concomitantly electrospinning two different polymer solutions. To this aim, the ES

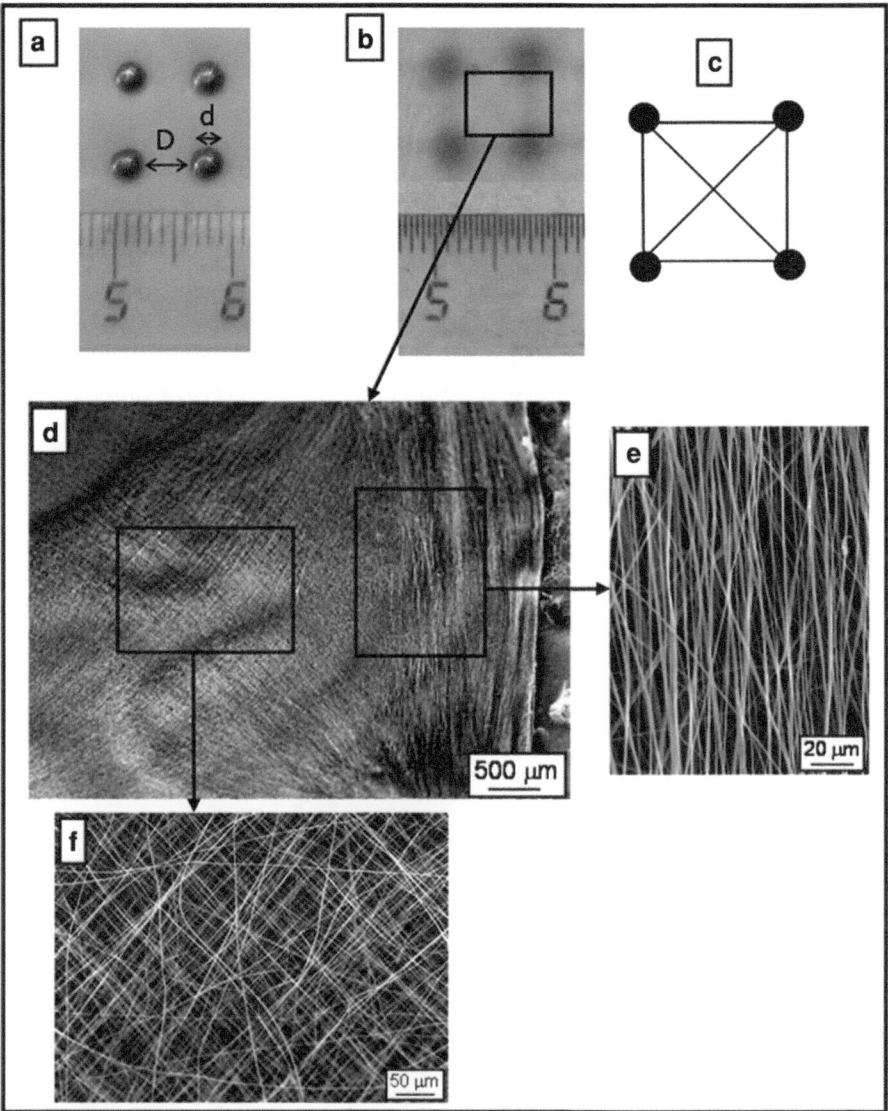

Fig. 3.36 a Pin array collector (d = 2 mm, D = 5 mm), **b** P(L)LA ES mat produced by using the pin array collector, **c** schematic representation of the macroscopic mat pattern, **d, e** and **f** SEM micrographs of differently oriented fibres at different mat positions

instrumental apparatus was modified by using two syringes connected to two distinct metallic needles, both positively charged, that were positioned on the opposite sides of the grounded cylindrical collector in order to address the ejected polymer jets on the same collector (see Fig. 3.37). The rotation of the cylinder ensures a homogenous distribution of the two different fibres on the entire mat.

A plane collector can be employed as well. In this case, however, the two needles are commonly vertically placed on the target that must translate in order to homogenously collect the fibres [81].

By using the instrumental apparatus sketched in Fig. 3.37, the following "composite" scaffolds were produced:

- scaffold made of a single polymer type but consisting of fibres with different morphologies. Figure 3.38a reports SEM micrographs of a P(L)LA scaffold composed of both micrometric and sub-micrometric fibres, as an example.
- scaffold made of two different polymers. Figure 3.38b shows a mixed scaffold consisting of P(L)LA fibres (thicker ones) and Poly(ethylene oxide) (PEO) fibres (thinner ones).

"Composite" ES scaffolds can be also composed by layers of different fibres obtained by consecutively electrospinning distinct polymer solutions on the same target [81, 82].

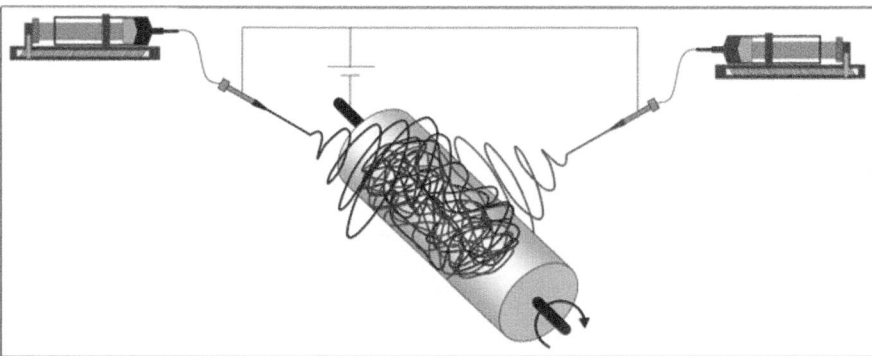

Fig. 3.37 Scheme of ES apparatus for the production of "composite" scaffolds

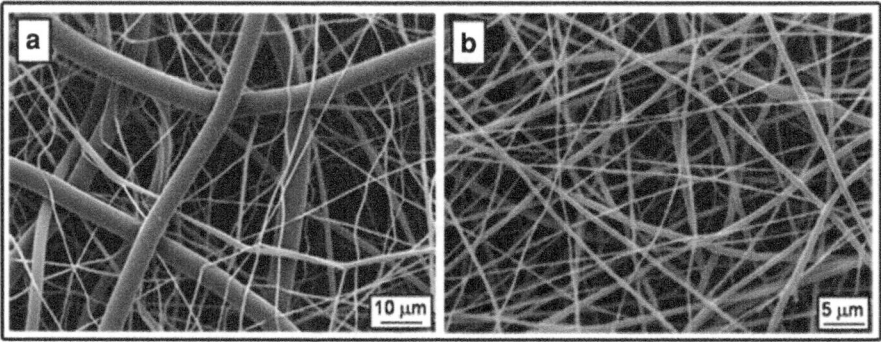

Fig. 3.38 **a** "composite" ES scaffold obtained by concomitantly electrospinning P(L)LA 13% w/V n DCM/DMF (65/35) and P(L)LA 20% w/V in CLF:DMF (90/10) ($\Delta V = 14$ kV, R = 0.02 ml/min, d = 20 cm); **b** "composite" ES scaffold obtained by concomitantly electrospinning P(L)LA 13% w/ V in DCM/DMF (65/35) and PEO 6% w/V in H_2O ($\Delta V = 18$ kV, R = 0.02 ml/min, d = 20 cm)

ES scaffolds consisting of different polymers and/or morphologies integrate, in the same product, the specific properties of each component. Thus, it is possible to design suitable products for the specific applications by combining fibres, that individually taken, do not display the desired mechanical properties, affinity towards cells and degradation rate.

3.2.6 Electrospun Scaffold Functionalization

3.2.6.1 Bulk Functionalization

One of the great advantages in using ES technique is the possibility to incorporate additives within fibres in a one-step process. This opportunity has been extensively exploited to load fibres with biomolecules that were intended to be released upon contact with the biological environment. Among the several drugs employed, antibiotics [83–85], anti-inflammatory drugs [86–88] and anti-cancer drugs [89] have been mainly investigated.

In the present research, given the key role of GFs in addressing cell behavior, an Endothelial Cell Growth Factor Supplement (ECGS) from Bovine Neural Tissue (Sigma–Aldrich) was chosen as bioactive additive. In the literature other types of GFs have been employed in combination with ES scaffolds. However, most of these studies concern the immobilization of such biomolecules at the fibre surface [90–92]. Leong et al. incorporated GFs that were intended to promote nerve regeneration into ES fibres of caprolactone-ethylethylene phosphate copolymer [93, 94]. In other studies, GF have been incorporated into fibres by using coaxial ES [95, 96].

In the present research, ES mats of $PLA_{50}GA_{50}$ were fabricated from a 15% w/V polymer solution in CLF/DMF/HFIP (60/30/10 by volume)[4] Sub-micrometric bead-free fibres with a diameter distribution of 790 ± 140 nm were obtained. ECGS in powder was added to similar polymer solutions in order to yield, upon solvent evaporation, ES $PLA_{50}GA_{50}$ mats containing two different amounts of ECGS: 0.4 and 4.8 wt%. The addition of such amounts of ECGS to the polymer solution does not remarkably affect fibre morphology in the ECGS-loaded scaffolds. ES mats supplemented with ECGS, as well as plain polymer mats not containing ECGS, were exposed to a culture of Mesenchymal Stem Cells, derived from human bone marrow, in order to evaluate the effect of GF on cell behavior and to assess if it retains its bioactivity once release in the culture medium [97]. The results of these experiments will be presented in "Cell culture experiments" of this chapter.

---

[4] ES process conditions: $\Delta V = 18$ kV, R = 0.02 ml/min, d = 12 cm, T = 22 ± 2 °C, RH = 45 ± 5%.

3.2.6.2 Surface Functionalization

Modification of biomaterial surface is commonly aimed at improving cell-biomaterial interaction and at reducing the negative effects of the foreign material in the organism. Therefore, one of the recent targets in scientific community is to change and to regulate the surface properties of ES fibres. The scientific literature investigating the possibility to modify ES fibre mats has been summarized in a recent review by Yoo et al. [98]. Several approaches have been followed, among them:

1. plasma modification: a proper selection of the plasma source enables to modify wettability and functional groups at the biomaterial surface in order to improve cell adhesion [99–101];
2. physical absorption of biomolecules: a variety of ECM components such as gelatin, collagen, laminin and fibronectin have been physically immobilized at ES fibre surface, generally after pre-treatment with plasma [102, 103];
3. surface graft polymerization: covalent immobilization of biomolecules has been performed at fibre surface after pre-treatement with plasma or UV radiation [104–107].

In the present work the surface of P(L)LA ES fibres was modified through an approach aiming at changing fibre properties by covering their surface with highly hydrophilic polymer brushes carrying the desired functional groups. The latter can subsequently covalently bind suitable biomolecules according to the specific biological response to be elicited. The hydrophilic nature of the polymer brushes should help controlling cell adhesion, by preventing non-specific protein absorption at the scaffold surface.

Investigation of cell-biomaterial interface has elucidated that cell adhesion to a surface is always mediated by a layer of proteins covering the biomaterial. On the contrary, a surface that does not absorb proteins cannot support cells, due to the absence of specific peptide sequences that allow cell attachment. Thermodynamic principles governing the adsorption of proteins onto surfaces involve a number of enthalpic and entropic terms that have been summarized by Ratner and Hoffman [108]. In brief, by limiting the discussion to neutral surfaces, protein-biomaterial interaction can be described as follows (see scheme in Fig. 3.39). In solution, proteins assume a globular structure, with hydrophobic groups at the core and hydrophilic residues in contact with water. When a hydrophobic biomaterial is exposed to a biological fluid containing proteins, the latter unfold and cover the surface in order to maximize hydrophobic interaction with the device. On the contrary, when the biomaterial is highly hydrophilic, protein absorption is an unfavourable process, thus cell attachment does not occur.

Despite the fact that protein absorption enables cell adhesion, non-specific absorption is in some cases undesired because it does not allow to control and regulate cell behaviour. In this context, the increase of wettability of ES fibres with hydrophilic polymer brushes is expected to prevent the occurrence of non-specific protein absorption, thus making the surface protein-repellent. At a later stage,

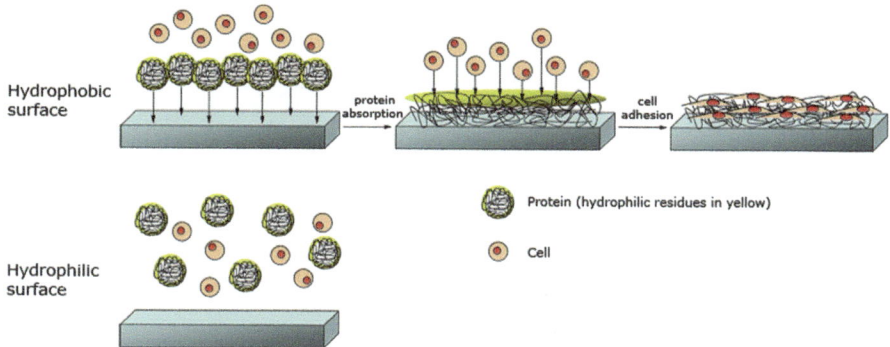

Fig. 3.39 Schematic representation of the process of cell attachment to a material surface (not drawn to scale). Hydrophobic surfaces absorb a layer of protein from the biological fluid and subsequently cells adhere to the surface thanks to the recognition of specific peptide sequences located along protein chains. Hydrophilic surfaces do not interact with the proteins in solution, thus cell attachment cannot occur

polymer brushes can be functionalized with biomolecules (e.g. specific peptide sequences) in order to specifically control cell attachment.

The present research preliminarily investigated the possibility of modifying ES fibres of P(L)LA. The work is based on a previous study carried out in collaboration with the University of Manchester, that investigated the functionalization of TCPS flat substrates with copolymer brushes of Glycerol Monomethacrylate (GMMA) and 2-(dimethylamino)ethyl methacrylate by Surface Initiated Atom Transfer Radical Polymerization (SI-ATRP) [109]. An analogous functionalization procedure was applied in this work to a 3D P(L)LA porous scaffold. The procedure was composed of three steps:

1. electrospinning of a P(L)LA solution containing star-branched oligo(D,L)lactic acid possessing carboxilate terminal groups (PLA-T6), with the aim of fabricating fibres with negatively charged surface upon water exposure;
2. physical absorption of a positively charged macroinitiator (MI, Fig. 3.40a) on fibre surface, with the aim of generating sites for initiating Atom Transfer Radical Polymerization (ATRP);
3. synthesis of polymer brushes by SI-ATRP of GMMA (Fig. 3.40b), with the aim of producing protein-repellent fibre surfaces.

ATRP belongs to the group of the controlled radical polymerizations (CRP). The main advantage of this kind of synthetic approaches is the possibility to polymerize a huge number of monomers obtaining polymers with wide ranges of molecular weights and low polydispersities. In CRP initiation reaction is fast, providing a constant concentration of growing polymer chains. In addition, termination reactions, responsible of the uncontrolled common radical polymerizations, are minimized. Termination reactions are almost completely suppressed by maintaining, during the polymerization, a very low concentration of reactive radicals.

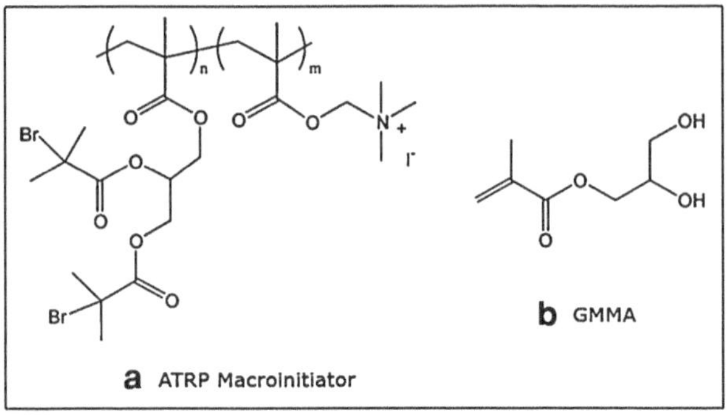

Fig. 3.40 Chemical
structures of **a** ATRP MI and
b GMMA

a ATRP Macroinitiator

b GMMA

This is obtained by using a radical species that is reversibly deactivated to a dormant species. It is worth noting that the equilibrium between the dormant and the active species is effective in controlling the polymerization if: (1) it is shifted towards the dormant species and (2) the rate of dormant-active species exchange is faster than the rate of propagation, so that all polymer chains have the same probability of adding monomer [110, 111].

In ATRP the dormant species is an alkyl halide (RX, R = Br, Cl, I) and its deactivation is promoted by the presence of a transition metal complex which undergoes an equilibrium between two oxidation states ($M_t^z–L_n \rightleftarrows X–M_t^{z+1}–L_n$). ATRP is widely employed as a versatile technique to prepare bioactive surfaces, including anti-fouling, antibacterial, stimuli responsive and patterned surfaces, through brushes grafting [112]. Fu et al. applied SI-ATRP for generating polymer brushes on fibres by electrospinning an ad hoc synthesized Br-terminated polystyrene [113]. Subsequently, ATRP was initiated by Br terminals present at fibre surface.

In the present research, the MI reported if Fig. 3.40a was absorbed on P(L)LA ES fibres through electrostatic interaction, with the aim to modify the fibre surface of commercial ES polymers. The MI should be easily and steadily absorbed on a negatively charged surface thanks to the cooperative effect of many positively charged repeating units. Moreover, the large amount of Br initiator groups carried by the MI is expected to generate densely packed polymer brushes. Negative charges can be generated on fibre surface by electrospinning a star-branched oligomer, possessing carboxylate terminals (PLA-T6), together with P(L)LA. The use of PLA-T6 oligomers should enable a good blending between the P(L)LA matrix and the additive. Moreover, it has been reported that, during fibre formation, the charge applied to the jet promotes the migration of the polarizable species contained in the solution (PLA-T6 in this case) towards the surface of the fibre [114, 115].

A solution of P(L)LA 13% w/V (DCM/DMF:65/35, by volume) containing 10% w/w of PLA-T6 was electrospun[5] and the obtained fibres had a average diameter of

---

[5] ES conditions: ΔV = 16 kV, R = 0.01 ml/min, d = 15 cm, T = 24 $\pm$ 2 °C, RH =33 $\pm$ 3%.

435 ± 125 nm. It is worth noting that by electrospinning the same P(L)LA solution not containing PLA-T6 additive, thicker fibres were produced (diameter distribution 610 ± 180 nm, see Sect. 3.2.2). It is well established that the addition of salts—in our case PLA-T6 oligomer carrying six carboxilate groups—to a polymer solution enables the obtainment of thinner fibres thanks to the increase of solution conductivity [43, 45, 116].

Scaffold specimens were immersed in a MI solution (0.1% w/V in water) overnight to allow electrostatic absorption to occur. The absorption conditions were selected according to previous results concerning absorption of the same MI on TCPS [109].

SI-ATRP of polyGMMA brushes was carried out on MI-coated scaffolds as described in "Materials and Methods". GMMA unit possesses two hydroxyl groups that confer hydrophilic character to the polymer brushes. Previous studies reported that GMMA is a non toxic, protein-repellent, and non cell-adhesive building block [109, 117, 118]. Moreover, its hydroxyl groups can act as reactive sites for subsequent desired functionalization reactions. Figure 3.41 compares fibre morphologies of P(L)LA scaffolds loaded with PLA-T6 at different stages of the functionalization procedure. Figure 3.41a reports SEM micrograph of the starting sample, Fig. 3.41b shows the scaffold after immersion in MI solution and Fig. 3.41c shows fibres after SI-ATRP synthesis of polyGMMA. Fibre surface was also explored at higher magnifications than that reported in Fig. 3.41 and no evident differences in fibre morphology were detected by SEM after the different treatments.

In order to verify the occurrence of MI absorption and of polyGMMA brushes growth at the fibre surface, electrokinetic analyses were performed. In general, when a charged solid surface comes into contact with a water solution, the free charges in solution distribute at the interface and they generate an electrochemical double layer. The latter is commonly divided into an immobile layer close to the solid surface and a mobile one. The potential at the interface of these two layers is called ζ-potential and its value is related to the charge density on the solid surface. ζ-potentials of the starting P(L)LA scaffolds, of samples after immersion in MI solution and of scaffolds after SI-ATRP were measured at pH values in the range 5–9 (see Fig. 3.42). Surface charge density of ES samples is highly affected by both MI and SI-ATRP treatments. At pH $= 6$ the ζ-potential of the starting sample was around -60 mV as a consequence of the presence of negative charges on the

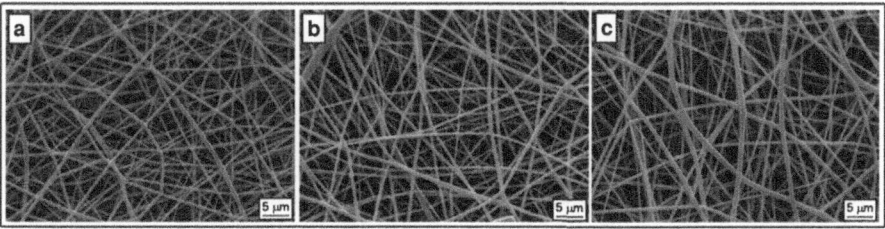

Fig. 3.41 SEM micrographs of P(L)LA fibres loaded with PLA-T6: **a** starting sample, **b** after MI treatment and **c** after SI-ATRP of GMMA

surface provided by the carboxylate groups of PLA-T6. After immersion in MI solution, the ζ-potential at pH = 6 increased to −20 mV. This finding can be explained by assuming that MI partially neutralizes the negative charges at the fibre surface, demonstrating that MI absorption occurs. After SI-ATRP the ζ-potential was 0 mV over the entire pH range investigated. Such a ζ-potential trend is explained assuming that a layer of neutral polyGMMA brushes covers fibre surface. Since hydroxyl groups in polyGMMA are very weak acids, the surface is neutral (ζ-potential = 0) over the entire range of pH investigated (pH = 5–9).

The successful modification of ES fibre surface was definitely confirmed by qualitative wettability tests. As shown in Fig. 3.43a, the starting sample is not wetted by water given both the intrinsic hydrophobic nature of P(L)LA and the presence of air entrapped within the pores of the scaffold. On the contrary, after the SI-ATRP procedure, the sample is instantaneously wetted by water (Fig. 3.43b).

Fig. 3.42 Electrokinetic data of ES samples differently treated

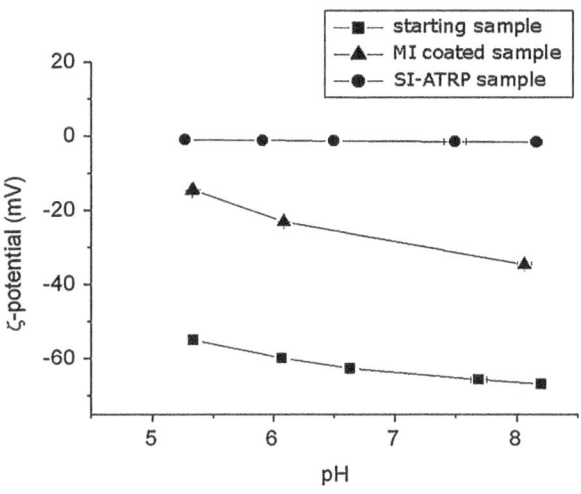

Fig. 3.43 Representative pictures of: **a** highly hydrophobic ES starting sample and **b** wet ES sample after polyGMMA SI-ATRP

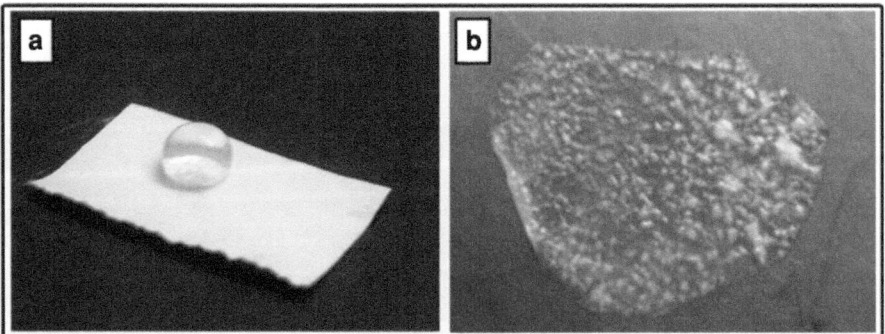

It is worth noting that mat 3D porous structure was not affected by the ATRP polymerization (see Fig. 3.41c), hence the change of mat wettability is ascribed to the presence of hydrophilic polyGMMA brushes covering fibre surface.

The preliminary results of fibre surface modification described above open a wide range of possibilities that, for a lack of time, haven't been attempted in the course of the present Ph.D. Indeed, the surface passivation towards protein absorption, and thus towards cell attachment, is the basic requirement to achieve a good control of cell behaviour. Future work will aim at exploiting the hydroxyl groups of polyGMMA brushes for immobilizing biomolecules, such as peptides or GFs, for eliciting the desired cell behaviour.

3.3 Hydrolytic Degradation Experiments

Scaffolds that are intended to maintain their role and functionalities for a limited period of time must inevitably undergo a process that leads to their gradual disappearance. Cell supports generally disappear as a consequence of a degradation process which cleaves the macromolecular chains. Sometimes these reactions are enzyme-catalyzed in vivo but usually polymer degradation process involves a simple hydrolytic cleavage of the chain bonds with a consequent reduction of chain length and of dissolution of water-soluble short molecules. The relative rates of water uptake and bond cleavage determine the degradation mode: when water uptake is faster than hydrolysis the process involves the entire sample ("bulk" degradation), when, on the contrary, water diffusion into the sample is slower than hydrolysis, surface erosion occurs. The first degradation mode is typical of polyesters while the latter is common to polyanhydrides [119]. Polyesters are the only class of materials employed in the course of the present research, therefore the following discussion strictly concerns the mechanism of "bulk" degradation.

To date, hydrolytic degradation studies carried out in vitro at physiological conditions have elucidated that polyester scission rate depends on a number of factors, e.g. chemical structure, morphology (amorphous to crystalline phase ratio) and specimen dimensions [120]. Chain hydrophilicity is the most important parameter affecting degradation. For poly(ω-hydroxyacids), hydrophlicity depends on the ratio between the number of ester groups and the number of carbon atoms along the chain. The rate of water uptake depends also on phase morphology: a densely packed crystalline phase is less permeable to water than the amorphous phase. Degradation can occur over several years for highly hydrophobic semi-crystalline polymers, such as PCL, or within few weeks for more hydrophilic amorphous materials, such as $PLA_{50}GA_{50}$. Another key parameter that affects the degradation process is specimen dimension which determines the occurrence of autocatalysis. Acidic autocatalysis is due to the presence of carboxylic end groups on chain fragments generated as a consequence of ester cleavage [120]. They decrease the pH in the surroundings with a consequent increase of hydrolysis rate.

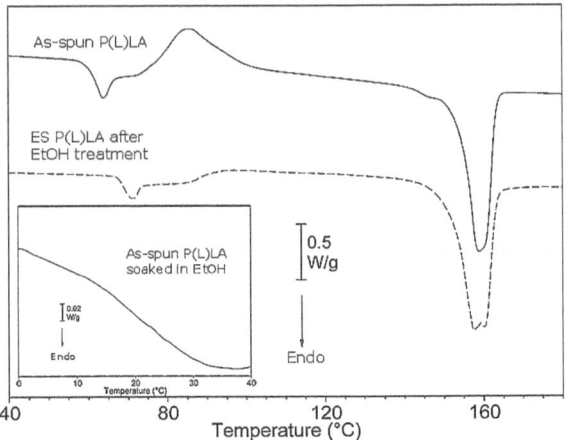

Fig. 3.44 DSC first scans of as-spun P(L)LA (*solid line*) and ES P(L)LA after EtOH treatment (*dotted line*). Insert: ES P(L)LA soaked in EtOH

Autocatalysis mostly affects the degradation rate of large specimens, owing to the limited diffusion of the acidic polymeric fragments that accumulate at the specimen core. In these cases, the degradation rate is faster in the bulk than at the sample surface.

Given the great importance of resorbability time-scale in the body, degradation mechanism of polyesters has been intensively studied in vitro, although, in the organism, other factors interfering with the degradation process can come into play. However, the in vitro experiments can help hypothesizing the kind of degradation mechanism and rate of a certain device in vivo.

In the present research, in vitro hydrolytic degradation experiments were performed on semicrystalline P(L)LA ES scaffolds. P(L)LA is universally recognized as a bioresorbable material and it is largely employed in biomedical applications. Its final degradation product is lactic acid, which is metabolized by the organism through the Krebs cycle.

Non-woven mats obtained by electrospinning a solution of P(L)LA 13% w/V in DCM/DMF (65/35 by volume)[6] were completely amorphous, as previously discussed in Sect. 3.2.2. In order to induce crystallization, as-spun fibres were immersed in EtOH. The alcohol plasticizes P(L)LA, whose T_g decreases from 60 °C down to 22 °C (by DSC analysis of as-spun P(L)LA soaked in EtOH, Fig. 3.44 insert). As a consequence, polymer crystallization window enlarges towards lower temperatures. In order to induce polymer crystallization, ES P(L)LA samples were therefore kept in EtOH overnight at RT. EtOH exposure did not induce any evident morphological changes to the ES fibres, which maintained their orientation and diameter distribution. Figure 3.44 shows DSC curve of ES P(L)LA after EtOH exposure overnight at RT together with DSC curve of the as-spun amorphous P(L)LA for comparison. In the latter the cold crystallization exotherm

---

[6] ES conditions: $\Delta V = 15$ kV, R = 0.02 ml/min, d = 15 cm, T = 22 ± 2 °C, RH = $41 \pm 3\%$.

peak is followed by a melting endotherm of the same entity. On the contrary, after EtOH treatment, DSC curve of ES P(L)LA displays a broad cold crystallization peak followed by a melting endotherm of higher entity, revealing the occurrence of crystallization upon EtOH exposure.

Hydrolytic degradation experiments on semicrystalline P(L)LA ES mats were run at 37 °C in buffer solution (pH = 7.4) up to a maximum of 396 days. After selected exposure times, the samples were recovered from solution and, after washing and drying, they were subjected to gravimetric, GPC, WAXS, DSC and SEM analyses. It is pointed out that buffer solution was periodically changed in order to maintain the samples at neutral pH, thus reproducing the physiological buffered environment. Samples are labelled X-P(L)LA, where X represents the days of permanence in buffer solution.

The GPC elugrams of ES samples recovered after selected hydrolytic degradation times are shown in Fig. 3.45, together with the GPC elugram of the undegraded sample (0-P(L)LA) as a reference. Over the whole time scale of the experiment, the P(L)LA macromolecules gradually eluted at higher retention volumes, thus indicating a progressive shift towards lower molecular weights. This result shows that the samples recovered from buffer after increasing exposure times contained a growing fraction of low molecular weight chains and a smaller number of long macromolecules. It is worth nothing that the elution curve, not only shifted towards lower molecular weights, but it also changed its shape. The initial elugram (sample 0-PLLA) had a strong peak corresponding to a molecular weight value of about 126,000 g/mol (green arrow in Fig. 3.45). The elugram of the starting sample also displayed a second weak peak in correspondence to a molecular weight of 6300 g/mol (red arrow in Fig. 3.45). In the course of degradation, the peak at higher molecular weights decreased and shifted towards higher elution volumes, while the peak at lower molecular weights gradually increased in the course of the

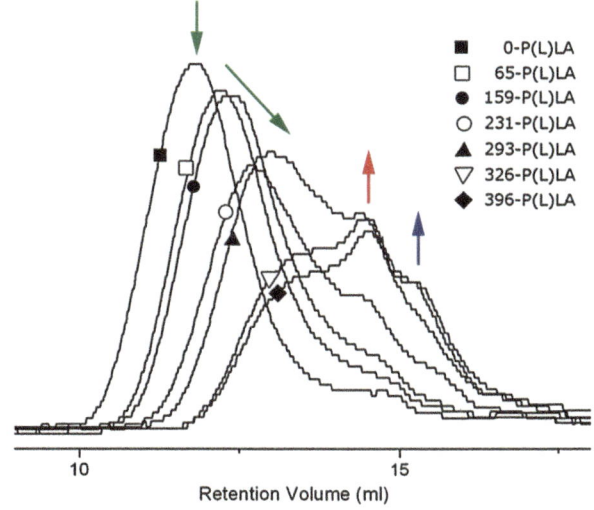

Fig. 3.45 GPC elugrams of P(L)LA mats retrieved from buffer after selected exposure time

■ 0-P(L)LA
□ 65-P(L)LA
● 159-P(L)LA
○ 231-P(L)LA
▲ 293-P(L)LA
▽ 326-P(L)LA
◆ 396-P(L)LA

Retention Volume (ml)

experiment, without changing its position. After about 290 days of buffer exposure, a strong third peak, corresponding to a molecular weight of 3100 g/mol, appeared (blue arrow in Fig. 3.45, sample 293-P(L)LA, 326-P(L)LA and 396-P(L)LA).

By integrating the GPC elugrams shown in Fig. 3.45, the molecular weight values reported in Table 3.5 were obtained. The Table lists both weight (M_w) and number average molecular weight (M_n) together with the percent of sample mass remaining after buffer exposure, m(%).

Figure 3.46 compares the changes of sample weight as a function of time with the corresponding mass changes at the molecular level (M_w changes), both quantities being expressed as percentage remaining after a given time in buffer solution. It is pointed out that, since the shape of the molecular weight distribution changed in the course of the degradation (see Fig. 3.45), M_w only provides an overall information of the change of polymer molecular weight as a consequence of hydrolysis. The molar mass significantly decreased from the very beginning of the degradation experiment, while the ES mats showed a very slow weight loss.

Table 3.5 Gravimetric and GPC results on hydrolytically degraded electrospun P(L)LA samples

Sample	Exposure time (days)	M_w (g/mol)[a]	M_n (g/mol)[a]	m (%)[b]
0-P(L)LA	0	172320	52340	100
65-P(L)LA	65	98310	14890	93.7
159-P(L)LA	159	81390	17680	91.9
231-P(L)LA	231	49860	8560	82.9
293-P(L)LA	293	32794	4975	80.1
326-P(L)LA	326	20600	3380	n.d.[c]
396-P(L)LA	396	19310	2740	n.d.[c]

[a] from GPC
[b] percentage weight remaining after buffer exposure, calculated according to Eq. 2.1
[c] n.d. = not detectable due to loss of sample integrity

Fig. 3.46 Weight percentage remaining (m%) (*filled circle*) and molar mass percentage remaining (M_w%) (*filled triangle*) as a function of buffer exposure time

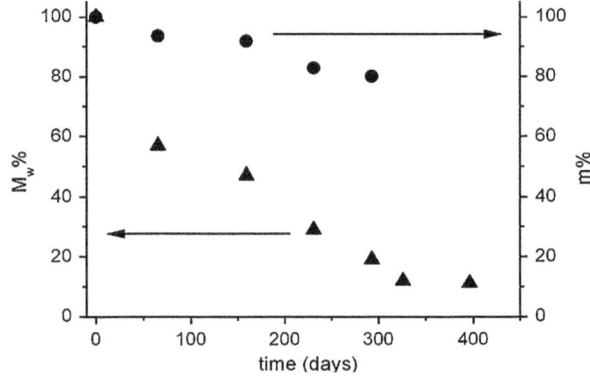

The observed weight loss is attributed to dissolution into the surrounding medium of low molecular weight chain fragments produced by hydrolytic degradation. It is pointed out that the weight loss was difficult to detect with the advancing of degradation. In the final part of the experiment (samples 326-P(L)LA and 396-P(L)LA) weight loss was not measurable anymore, due to sample fragility and loss of integrity. It is interesting to note that around day 290 the rate of molecular weight decrease abruptly slowed down. This result can reflect the absence, in samples recovered after more than 290 days in buffer, of the short chain fragments produced by hydrolysis, which were responsible of the fast M_w decrease up to day 290. For exposure times longer than 290 days, such fragments were likely released from the sample via water solubilization.

WAXS analysis were performed on P(L)LA ES samples recovered after different buffer exposure times, in order to evaluate changes in the crystal phase during degradation. Figure 3.47a reports diffractograms of selected samples. The two crystalline reflection peaks at $2\theta = 16.5°$ and $19°$ are typical of the α-form of P(L)LA having a pseudo-orthorhombic unit cell [121] and their intensity increased with the progress of degradation. Figure 3.47b plots the crystallinity degree—calculated as the ratio of the crystalline peak areas to the total area under the scattering curve—as a function of degradation time and shows that the degree of crystal phase sharply increased after around 230 days. It is pointed out that, at a similar time scale of degradation, the decrease of average molecular weight, previously reported in Fig. 3.46, slowed down as a consequence of low molecular weight chains dissolution. Therefore, it is reasonable to assume that degradation mostly occurred in the amorphous phase and that the amount of crystal phase mainly increased as a consequence of macromolecules released from the amorphous phase. However, crystallization of part of amorphous molecules during buffer exposure can not be excluded.

DSC curves of selected P(L)LA samples after different exposure times are reported in Fig. 3.48. The single melting peak at $T_m = 160$ °C (sample 0-P(L)LA)

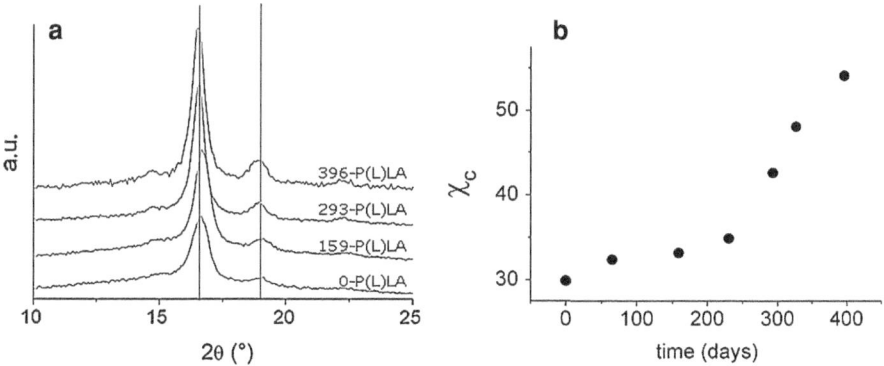

Fig. 3.47 WAXS data: **a** diffractograms of selected P(L)LA samples retrieved from buffer after selected exposure times and **b** change of crystallinity degree in the course of degradation experiment

Fig. 3.48 DSC first scans of selected P(L)LA samples after different times of buffer exposure (heating rate 20 °C/min)

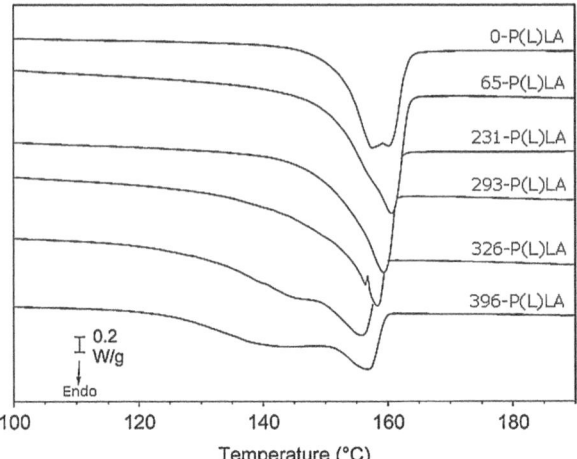

shifted and broadened toward lowers temperature with the advancing of degradation (samples 65-P(L)LA, 231-P(L)LA and 293-P(L)LA). Interestingly, last samples retrieved from the buffer (i.e. samples 326-P(L)LA and 396-P(L)LA) displayed a melting endotherm characterized by an additional peak at low temperatures ($T_m = 145$ °C).

SEM observations of selected ES mats retrieved during the degradation experiment show a gradual change of fibre morphology (Fig. 3.49). After 230 days of degradation, fibres appeared broken in some points (Fig. 3.49b) and the last two samples retrieved from the buffer (326-P(L)LA and 396-P(L)LA) showed many broken fibres. Moreover, the initial smooth fibre surface (see Fig. 3.49a) became rough as a consequence of the degradation (see inserts in Fig. 3.49c, d). As previously discussed, the measured of weight loss was impossible for samples 326-P(L)LA and 396-P(L)LA due to their fragility and fragmentation (see m% values in Table 3.6). Therefore, it cannot be excluded that fibres shown in Fig. 3.49c and d might have been broken during preparation of the sample for SEM analysis.

Gravimetric, GPC, WAXS, DSC and SEM analyses, taken together, provide information about the hydrolysis of semicrystalline P(L)LA ES fibres. As previously mentioned, the slowdown of the decrease of M_w occurring around 290 days of degradation (by GPC, see Fig. 3.46), coupled with the increase of crystallinity degree at a similar time scale (by WAXS, see Fig. 3.47), revealed that hydrolysis mostly occurred in the amorphous phase. Particularly significant is the change of molecular weight distribution in the course of degradation. The shape of GPC elugrams reported in Fig. 3.45 can be explained referring to Vert's thorough studies concerning hydrolytic degradation of polylactide thick films. Vert et al. ascribed the formation of multimodal molecular weight distributions to the presence of macromolecular chains degrading with a different rate in the course of the experiments, i.e. (1) macromolecules in amorphous phase that degrade faster than macromolecules in the crystal phases and/or (2) macromolecules in specimen core which degrade faster than those at the surface, as a consequence of the autocatalytic effect

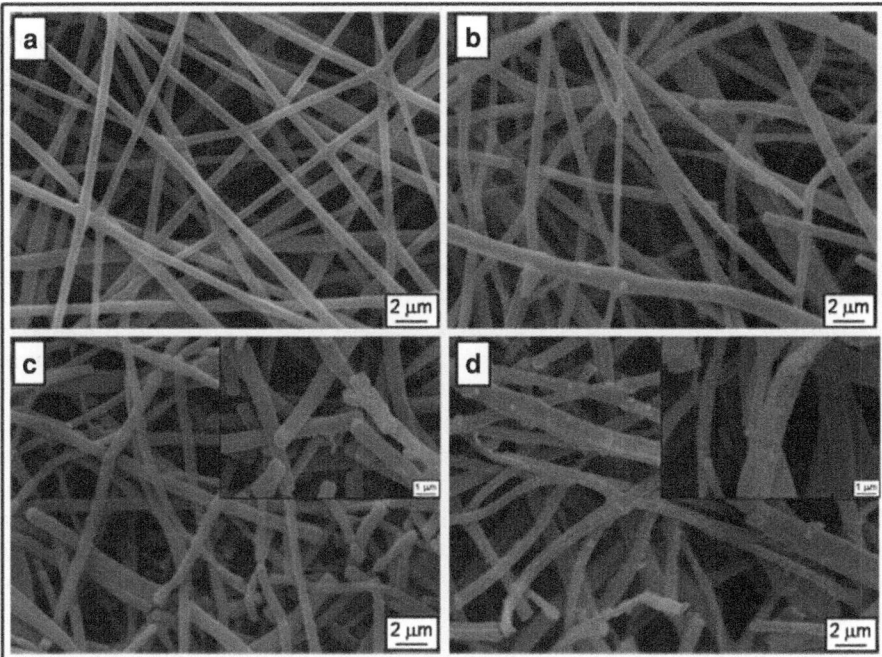

Fig. 3.49 SEM micrographs of: **a** 0-P(L)LA, **b** 231-P(L)LA, **c** 326-P(L)LA and **d** 396-P(L)LA

[122–124]. When the authors studied thick specimens of completely amorphous polymers, the molecular weight distributions displayed two distinct peaks, only as a consequence of the autocatalysis. In the case of semicrystalline thick films, the autocatalytic effect combined with the different degradation rates of amorphous and crystalline phase led to multi-peak molecular weight distributions. In the case of ES materials, the autocatalysis is unlikely given the extremely small fibre section. This point was confirmed by a previous study investigating the hydrolytic degradation of completely amorphous $PLA_{50}GA_{50}$ ES fibres [97].

The progressive change of semicrystalline ES P(L)LA molecular weight distribution (see Fig. 3.45) in the course of degradation experiment might be interpreted as follows. At the initial stages degradation interests macromolecules in the amorphous phase. The peak corresponding to the high molecular weight fraction ($M_W = 126,000$ g/mol, green arrow in Fig. 3.45) shifted towards lower molecular weight values and it decreased, while the low molecular weight fraction peak increased ($M_W = 6300$ g/mol, red arrow in Fig. 3.45). The latter did not change its position probably because the molecules produced by hydrolysis having molecular weights lower than 6300 g/mol dissolved in the buffered medium, thus they were not detectable by GPC analysis of the solid recovered sample. With the proceeding of degradation, after 290 days of buffer exposure, the appearance of a further peak at higher elution times ($M_W = 3100$ g/mol, blue arrow in Fig. 3.46) reveals the presence of very low molecular weight fragments which cannot dissolve in water. The appearance of this third peak in the GPC elugram can be explain as follows. Water does not penetrate the densely packed crystal phase, but it has access to crystal surface, where it can cleave ester bonds located on folded amorphous chain portions. However, if other portions of the same macromolecule are blocked in the crystal phase, the chain cannot dissolve in the aqueous environment. Therefore, these low molecular weight fragments blocked in the crystal phase can be detected by GPC analysis of the recovered solid samples. This hypothesis might also explain DSC results previously reported in Fig. 3.35. Indeed, with the proceeding of degradation, DSC curves showed the appearance, in the melting endotherm, of an additional peak at low temperatures ($T_m = 145\ ^\circ C$) which may be associated with the melting of short chain fragments.

This study provides data on the time-scale of in vitro hydrolytic degradation of semicrystalline ES P(L)LA mats. Results demonstrated that, despite the polymer molecular weight started to decrease from the very beginning of the experiment, weight loss was very slow within 10 months (290 days) of buffer exposure. After this period of time, the sample became fragile and loss its integrity, thus loosing its structural functionality. This can lead to the release of small pieces of fibres in the body which can be phagocytized by foreign body-giant cells [125, 126]. Such kind of information is crucial in the design of scaffolds for TE that should maintain their mechanical performance for a given period of time after which they should be resorbed.

3.4 Cell Culture Experiments

In the course of the present Ph.D., ES scaffolds, designed, fabricated and characterized as described in the previous chapters, were employed in cell culture experiments performed in collaboration with biochemical laboratories. The contribution of material science was precious in the context of these interdisciplinary collaborations. Indeed, a deep knowledge of polymer chemical-physical properties is a necessary requirement, not only to design and fabricate ES products, but also to improve standard cell culture operating procedures in order to perform biological experiments by using unconventional cell substrates, i.e. ES scaffolds.

Cell culture experiments illustrated in this chapter were carried out in collaboration with the Biochemistry Department "G. Moruzzi" (University of Bologna), the Clinical Department of Radiological and Histocytopathological Sciences (University of Bologna) and the Foundation for Development of Cardiac Surgery (Zabrze, Poland).

3.4.1 Scaffold Preparation for Cell Culture Experiments

Prior to any cell culture experiment, scaffold sterilization is a necessary step for preventing bacterial and fungi contamination and it is even more important for hindering the transmission of infectious pathogens to patients when implantable devices are used. Common biomaterial sterilization techniques are: (1) treatment with ethylene oxide (EtO) vapour, (2) thermal treatment, (3) irradiation with UV light or (4) with γ-rays.

EtO treatment is highly recommended for biomaterial sterilization. However, this substance is highly toxic and inflammable, thus sterilization is often performed by specialized companies at high costs. Moreover, since EtO sterilization is often performed around 50–60 °C, its use is not recommended for polymers with T_g values in this temperature range (e.g. $PLA_{50}GA_{50}$).

Thermal treatment is usually carried out by gradually increasing water vapour temperature up to approximately 120 °C. While these conditions grant for the death of many bacteria, they can be detrimental for polymers that melt below 120 °C or that are susceptible of hydrolytic degradation.

Other sterilization procedures involve irradiation. Among them, the most accessible method consists in using ultraviolet (UV) light which sterilizes without inducing secondary reactions. However, the low penetration ability of UV radiation makes the efficacy of this technique limited, especially for thick fibre mats. Compared to UV light, γ-rays offer a more complete sterilization since they are able to penetrate into the irradiated materials. Possible concerns arise from alterations of the irradiated polymer which can undergo chain cleavage, cross-linking reactions or oxidation.

The most commonly used method for sterilizing ES scaffolds is Ethanol (EtOH) wetting and its effectiveness has been largely demonstrated. It is pointed out, however, that EtOH is not considered as a sterilizer but it is classified as a disinfectant, being effective against bacteria, fungi and many viruses but not against bacterial spores. ES meshes are simply immersed in EtOH for 15–30 min and extensively washed before use, typically with phosphate buffered saline solution or cell culture medium added with antibiotics. A great advantage in using EtOH is that it allows to wet the intrinsically hydrophobic mats because it spontaneously enters the pores of the structure. However, EtOH sterilization also has a drawback because some ES mats are not dimensionally stable when immersed in EtOH. Such scaffolds may undergo a macroscopic shrinkage that is accompanied by microscopic changes of fibre morphology: fibres become curly, fibre diameter increases and pore size decreases. This behaviour has been reported for some polymers such as $PLA_{50}GA_{50}$ or P(D,L)LA, whereas other materials such as semicrystalline PCL do not undergo extensive shrinkage in the same experimental conditions [83, 127, 128].

In ES scaffolds, the macroscopic shrinkage and the change of fibre morphology has been attributed to changes of molecular conformation due to chain relaxation occurring when macromolecules in the amorphous state acquire mobility [128]. Indeed, when the fibre is generated during the ES process, polymer chains are stretched in the fibre axis direction, while the solvent quickly evaporates. If the stretched molecular chains do not have enough time to undergo relaxation before complete solvent evaporation, they will solidify in an elongated conformation. Afterwards, if for any reason macromolecules acquire mobility at a temperature close to or higher than their T_g, they will change their conformation from the stretched oriented one towards the thermodynamically stable random coil one. EtOH acts as a plasticizer for many of the polyesters commonly employed in TE. Therefore, if the presence of EtOH leads to a decrease of polymer T_g below RT, the shrinkage can occur during the sterilization procedure at RT.

The effect of EtOH on fibre morphology is clarified by the following example. A square ES mat (30 × 30 mm) of P(LA-TMC) was placed in EtOH for 1 h at room temperature. The T_g of ES P(LA-TMC) is around 34 °C and it decreases down to 0 °C when the mat is immersed in EtOH (by DSC analysis of as-spun P(LA-TMC) fibres soaked in EtOH, DSC curve not shown). Figure 3.50 shows the effect of EtOH on random P(LA-TMC) fibres together with the percentage of shrinkage calculated as:

$$s = \frac{\ell - \ell_0}{\ell_0} \times 100 \qquad (3.1)$$

Where ℓ is the side length of the mat after EtOH treatment and ℓ_0 is the initial side length (30 mm). After 1 h in EtOH, fibre morphology changed: fibre diameter increased, fibres became more packed and pore dimension decreased (compare Fig. 3.50a with b). As already pointed out, this finding can be attributed to chain relaxation occurring during EtOH treatment. Moreover, when patterned mats or mats made of aligned fibres are placed in EtOH they loose completely their original fibre orientation. These changes of scaffold dimension and of fibre morphology introduce obvious limitations in the use of those ES materials that undergo strong shrinkage during their sterilization and some authors have even excluded the use of such polymers for TE applications [83, 127, 128].

This drawback has been circumvented in the course of this research by a procedure that allowed the macromolecules to relax, while preventing shrinkage. With this aim in mind, the ES scaffold was attached to a rigid plastic frame and placed in EtOH. Being the mat bound to the frame, its gross dimensions did not change and fibres tended to maintain their morphology (compare Fig. 3.50a with c). After this "constrained" pre-treatment in EtOH, the scaffold was removed from the frame and it was immersed again in EtOH to ascertain whether any dimensional changes occurred in this second wetting step. Figure 3.50d shows that, although some very limited shrinkage still occurred, fibre morphology was maintained even if the scaffold was not constrained anymore.

As already pointed out, when P(LA-TMC) mats are immersed in EtOH, macromolecules undergo transition from the frozen, glassy state to the mobile state. Therefore, results obtained using the above described "constrained" treatment may be interpreted as follows. During the "free" EtOH treatment (i.e. when the scaffold was not "constrained" by the rigid frame) macromolecules relaxed from the stretched conformation and underwent spontaneous coiling. A change of fibre morphology and mat dimensions was therefore observed (Fig. 3.50b). These changes did not occur when the mat was immersed in EtOH after fixing it to a rigid frame ("constrained" treatment) (Fig. 3.50c). It should be pointed out that also in this case chain relaxation phenomenon occurred (i.e. macromolecules partially changed their conformation from the aligned to the entropically favoured coiled

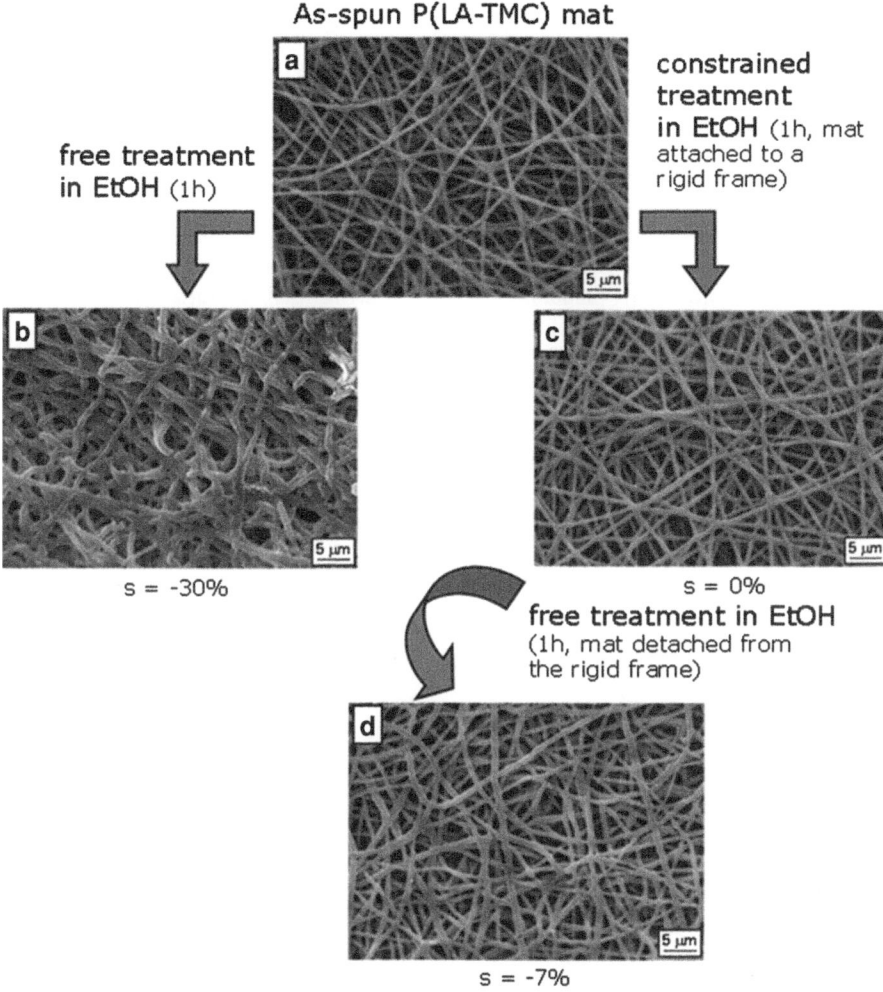

Fig. 3.50 Effect of EtOH treatments on P(LA-TMC) fibre morphology. s is the percentage of shrinkage calculated according to Eq. 3.1

one) but chains also tended to flow one respect to the other, since the constrained fibres had fixed length and they could not macroscopically follow the change of molecular conformation. Afterwards, when the scaffold was immersed again in EtOH, without any constraint, only some residual shrinkage was observed (Fig. 3.50d), that is attributed to a minor fraction of chains still in the stretched oriented conformation. It is reasonable to assume that a longer "constrained" treatment in EtOH would have allowed to completely eliminate any residual shrinkage. In conclusion, the "constrained" treatment performed before scaffold sterilization in EtOH can be an effective way to limit or totally eliminate scaffold shrinkage and fibre morphology changes and may broaden the range of ES polymers that can be employed for TE applications, using EtOH sterilization.

In this work, with the aim to prevent scaffold shrinkage and to allow an easy handling of ES scaffolds during cell culture experiments, ES mats were therefore attached to plastic rings (Tecaflon PVDF) by using non-toxic silicone (see Materials and Methods). A similar commercial solution is alternatively offered by Scaffdex, a company that sells "crown rings" to immobilise thin membranes. These inserts are typically custom-made in order to fit in common culture well-plates with given dimensions (see Materials and Methods). Both solutions allow to easily handle ES mats which are typically thin and consist of extremely light fibrous networks that easily twist, wind up, or fold. Scaffold fixation on plastic rings also avoids cell dispersion/outflow during cell culture experiments by obtaining a cell-leakage-proof well that can be employed to carry out quantitative cell culture experiments.

3.4.2 Characterization of Cells Cultured on Electrospun Scaffolds

According to TE approach, initially cell culture is carried out in vitro and the implantation of the cell-scaffold construct comes after a variety of cell assays aiming at characterizing and identifying the behaviour of cells in contact with the synthetic scaffold. To this aim, scientific community has successfully applied some of 2D cell culture characterization tests to 3D cell cultures, without any substantial modification of the protocols. However, it is sometimes impossible to characterize cells without adapting the specific assay to the specific scaffold used.

To date, the behaviour of cells cultured on ES scaffolds is investigated mainly through the following techniques:

- MTT, MTS or Alamar Blue colorimetric assays for the evaluation of cell viability and proliferation [39, 129–131]. They exploit the change of absorption spectra of a chromophore as a consequence of its reduction by cell reductase enzymes. Since reduction takes place only when enzymes are active, these tests measure the metabolic activity of viable cells which provides an indirect information on cell number.
- Real Time Polymerase Chain Reduction (PCR) for determining the abundance of different RNA molecules within the cells in order to assess regulation of gene expression [130, 132–134].
- Western Blotting for semiquantative detection of specific protein expression [132, 134, 135]. After cell detachment from ES scaffold, proteins are extracted from cells and they are separated by using gel elecrophoresis.
- SEM observations which represent the gold standard for the evaluation of cell morphology.

Histochemical and immunohistochemical techniques, which are normally employed for the investigation of cells in native tissues, are not largely used for the characterization of in vitro cell cultures on 3D scaffolds, despite they provide unique information to identify cell behaviour when analysing native tissues. These techniques consist in observing thin slides of tissue under light microscope. Staining of nuclei and cytoplasm (histochemistry) or of proteins, carbohydrates and lipids (immunohistochemistry) with specific antibodies enables to visually investigate cell behaviour and metabolism. When these techniques are applied to the study of cells grown in vitro on 3D scaffolds, the preparation of the sample and the overwhelming difficulties in obtaining good sections of the porous structure have limited their use. Indeed, it has been reported that aliphatic polyesters (the main constituent materials of bioresorbable scaffolds) are very sensitive to common processing for the preparation of the sample for histo-chemical and immunohistochemical analysis, which is believed to lead to scaffold artifactual modifications [136–138]. As a matter of fact, common protocols involve the use of temperatures higher than RT and organic substances which might be solvents for the scaffolds.

The cryosection procedure is a less destructive protocol largely applied in the literature for obtaining cross-cryosections of ES samples without using organic solvents. Histochemical staining of these cross-sections gives the possibility to study cell infiltration into the porous structure which is a key issue for the inte-gration of the implanted material into the recipient tissue [132, 139, 140]. How-ever, this procedure does not provide optimal sections for histochemical and immunohistochemical analysis since it often leads to scaffold fragmentation and loss of cell details.

In order to circumvent the above discussed drawbacks, information similar to that provided by immunoistochemistry can be obtained by immunofluorescence performed directly on nanofibrous scaffolds [131, 133, 134, 141]. In this case the antibody is tagged with a fluorophore which is visualized under the UV-light microscope. Immunofluorescence enables evaluation of specific antigen expres-sion, distribution of focal adhesion contacts and of cell differentiation markers. However, major drawbacks are low spatial resolution and polymer autofluores-cence, hence the highly expensive confocal microscopy is often needed to analyze these 3D samples [39, 129, 130].

The limitations of the above described methods suggest that evaluation of cell nanofibre interactions is still technically demanding. In the course of the present Ph.D., knowledge of material and scaffold properties were fundamental in order to correctly work with ES samples seeded with cells. Besides most of the common analysis already employed in the literature for characterizing cells seeded on ES scaffolds (e.g. SEM, Alamar Blue assay, Western blotting, immunofluorescence, etc.), a new sample preparation procedure, commonly used for native tissues, was adapted in order to perform histological and immunohistochemistry analysis on ES samples. The protocol, applied to P(L)LA ES fibres, was based on automatized and standardized Formalin Fixed Paraffin Embedding (FFPE) procedure. A processing temperature around 56 °C and the use of suitable organic substances that does not dissolve P(L)LA enabled to successfully embed the scaffold in a paraffin block. Both cross-sections and "en face" slices were obtained and they were subsequently stained for conventional histological and immunohistochemical analysis [142]. The applied protocol allowed the preparation of intact sections of P(L)LA ES fibres without damaging the structural integrity of the scaffold. Indeed, P(L)LA has a glass transition temperature around 61 °C, thus higher than the temperature used during the treatment with melt paraffin (T = 56 °C). It should be taken into account that polymeric biomaterials possessing transition temperatures lower than embedding process temperature might not be successfully processed using the same protocol.

The preparation of cross and en face-sections of ES samples seeded with cells opens the possibility to easily obtain qualitative and semi-quantitative information—e.g. cell morphology, ECM production, cell adhesion, death and immunophenotype expression—essential whenever a bioresorbable scaffold is designed for in vivo applications. Moreover, FFPE cross-sections offers the possibility to analyze cell infiltration providing clearer cellular details than the more common employed cross-cryosections.

In this work, histochemical and immunohistochemical analysis, together with the other cell characterization techniques (e.g. SEM, Alamar Blue assays, immunofluorescence, etc.), were applied in the course of the cell culture experiments in order to evaluate cell behaviour and cell-scaffold interactions.

3.4.3 Biocompatibility Evaluation of Electrospun Scaffolds: The Example of PPDL

As previously discussed in Chap. 1, the primary requirement of any biomedical device is to be biocompatible. Therefore, whenever a new device is used in contact with a biological environment, biocompatibility is imperatively the first characteristic to be verified. Unfortunately, biocompatibility is not measurable and a clear operative definition of it does not exist. However, some tests and procedures have been development to evaluate biocompatibility. Commonly, in vitro tests are firstly carried out, followed by in vivo animal tests and clinical trials. According to the particular application for which the device has been designed, different kinds of tests can be performed, such as cytotoxicity, mutagenesis, emocompatibility, etc.

These general considerations must be applied also to new ES materials aiming at supporting cell growth for inducing tissue regeneration. In the course of this Ph.D. the collaboration with biological laboratories enabled to carry out in vitro tests to evaluate biocompatibility of newly synthesized polymers and copolymers. As an example, biocompatibility studies of ES poly(ω-pentadecalactone) scaffolds will be presented in this paragraph. PPDL is a non-commercial polymer which is synthesized at present only at laboratory scale. The PPDL used in the course of the present research activity was synthesized by enzyme catalyzed polymerization and it was ES by using organic solvents (CLF, DCM and HFIP, see Sect. 3.2.2) which are highly toxic towards cells. Despite the fact that TGA analysis confirmed the absence of detectable residual solvents, it is important to asses the effect of solvent traces eventually contained in the ES fibres. The biocompatibility of PPDL ES scaffolds towards mammalian cells was evaluated by using the H9c2 cell line (myoblast cells derived from embryonic rat heart). These cells were used as in vitro benchmark to test indirect cytotoxicity as well as cell adhesion, proliferation and morphology of cells seeded on PPDL fibres[7] [143].

Indirect cytotoxicity was firstly investigated. This kind of test evaluates harmful effects on mammalian cell cultures arising from the release of toxic substances from the device. Indeed, all materials can release, to a different extent, low molecular weight substances such as residual monomers, additives, impurities or residual processing solvents (in the case of ES materials). Indirect cytotoxicity evaluation of ES PPDL scaffolds was performed in accordance with the ISO10993-5 international standard for biological evaluation of medical devices.[8] In brief, ES PPDL was kept in Dulbecco's modified Eagle's Medium (DMEM) in order to extract the low molecular weight substances from the fibres. The DMEM containing the PPDL extracted substances was added to a culture of H9c2 cells and viable cells were quantified by using the sulforhodamine B (SRB) colorimetric assay for cytotoxicity screening.[9] The result was compared both with the negative control (cells culture in DMEM not kept in contact with ES PPDL) and with the positive cytotoxic control (cells culture in DMEM added with 1 mM H_2O_2). Figure 3.51 shows that SRB absorption spectroscopy output was comparable for samples grown for 24 h in PPDL-extraction medium or in standard DMEM medium whereas, when exposed to 1 mM H_2O_2, as a positive cytotoxicity control, all the cells died (not shown). This result indicates the absence of potentially cytotoxic products released from PPDL, in agreement with the conclusion of recent indirect cytotoxicity tests [144] run on mouse fibroblast 3T3 cells treated with an extract from a PPDL bulk sample. In addition, we demonstrated that PPDL non-cytotoxicity was maintained also after scaffold fabrication via electrospinning by using CLF, DCM and HFIP as solvents.

Cytocompatibility tests were also carried out by seeding cells directly on PPDL ES scaffolds.[10] In order to evaluate cell adhesion and proliferation on the electrospun scaffold, the number of viable cells was quantified every other day, up to

---

[7] A PPDL solution (7% w/V) in CLF/DCM/HFIP (50/40/10, by volume) was electrospun using the following process conditions: $\Delta V = 16$ kV, R = 0.05 ml/min, d = 15 cm, T = 23 $\pm$ 2 °C, RH = 40 $\pm$ 5%.

[8] PPDL mats were immersed in DMEM (5 mg polymer/1 ml medium) supplemented with 10% heat-inactivated fetal bovine serum (FBS), 2 mM L-glutamine and 100 U/ml pen/strep, at 37 °C in a humidified atmosphere containing 5% CO_2 for 24 h in order to obtain the medium containing the PPDL (PPDL-extract medium). H9c2 cells were seeded in a 96-well culture plate (500 cells/well) in standard DMEM to allow their attachment. After 48 h, the culture medium was discarded, the PPDL-extract medium was added to the wells and the cells were further incubated for 24 h.

[9] Optical density ($\lambda = 540$ nm) of samples was read in a Wallac VICTOR multilabel multiplate reader (Perkin Elmer). Two separate experiments, six replicates each, were performed. Optical density mean values $\pm$ standard error of the mean (sem) for replicates were calculated and the unpaired t-test was used to evaluate statistical differences between mean values.

[10] 2.5×10^4 H9c2 cells in 1 ml DMEM were seeded onto the PPDL mat.

Fig. 3.51 Evaluation of the indirect cytotoxicity of ES PPDL scaffolds. Cell viability in PPDL extraction media was not statistically different than that in DMEM (Reprinted from [143], Copyright (2010), with permission from Brill Publisher)

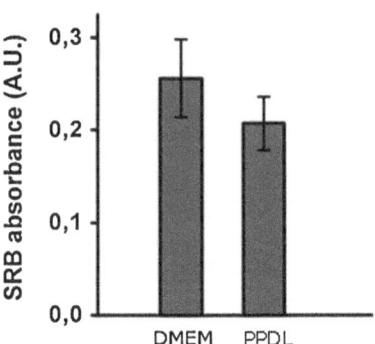

14 days, with the Alamar Blue fluorescence assay.[11] The same cell culture was monitored over time, thus overcoming drawbacks inherent to other sample-destructive proliferation assays, such as the MTT and MTS tests. Control signal was acquired from H9c2 cells cultured in TCPS. Results reported in Fig. 3.52 show that, after 24 h from cell seeding, ES PPDL mats hosted about 50% of the number of H9c2 cells that adhered to the control polystyrene surface (day 1). The number of cells growing onto PPDL increased linearly for up to day 14 (end of experiments), although a significant difference with respect to the number of the cells proliferating onto the TCPS was maintained at any given time point ($p < 0.0001$ by ANOVA). Alamar Blue assay showed that PPDL fibrous substrates are non cytotoxic towards H9c2 cells and support cell proliferation. Moreover, when let to grow up to 4 weeks, H9c2 cells reached the same maximum value obtained on the polystyrene support (data not shown).

The morphology of cells grown onto ES PPDL mats was observed by SEM. Figure 3.53a shows that, after 14 days of culture, H9c2 cells spread over the PPDL mat surface while retaining their native, mesenchymal spindle-shaped, sheet-like morphology. Furthermore, at 14 days, the scaffold surface was almost entirely covered by cells. In the experiment where cells were allowed to grow over PPDL for up to 4 weeks, the PPDL surface appeared completely covered with a cell monolayer that prevented visualization of underlying fibres (Fig. 3.53b).

SEM observation of cell morphology together with results provided by indirect cytotoxicity test and Alamar Blue assay confirmed that PPDL in the form of ES fibres is biocompatible and is able to promote H9c2 adhesion and proliferation, thus being a promising slow-degrading support for TE applications.

---

[11] Blue fluorescence (Ex/Em = 540/590 nm) was read in a Wallac Victor multilabel multiplate reader (Perkin Elmer). Four separate experiments ($n = 4$), three replicates each, were performed. Two-way analysis of variance (ANOVA) was performed to compare proliferation curves. Values were given as the mean values of fluorescence $\pm$ sem.

3.4.4 Electrospun Fibres Loaded with Growth Factor: Effect on Stem Cell Culture

GFs are proteins dissolved in the gel-like component of ECM and have a key role in driving cell activity. They are frequently added to cell culture media during common in vitro cell culture experiments in order to induce cell proliferation and differentiation. Given their biological importance, several efforts are made in TE for achieving a gradually release of GFs from the scaffold upon contact with physiological fluids.

In the course of this Ph.D. an Endothelial Cell Growth Factor Supplement (ECGS) from Bovine Neural Tissue (Sigma–Aldrich) was incorporated within ES fibres through a one-step process. $PLA_{50}GA_{50}$ ES scaffolds loaded with different amounts of ECGS were prepared as previously described in Sect. 3.2.6.1. ECGS is typically formulated to stimulate cell proliferation activity and to induce stem cell

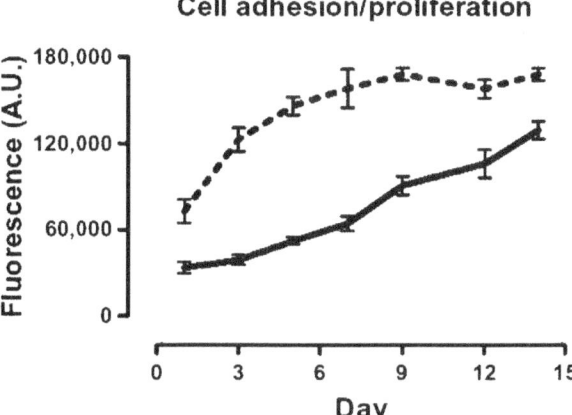

Fig. 3.52 Evaluation of cell adhesion and proliferation on the ES PPDL scaffolds (*continuous line*) compared with TCPS control (*dotted line*) by Alamar Blue fluorescence assay (Reprinted from [143], Copyright (2010), with permission from Brill Publisher)

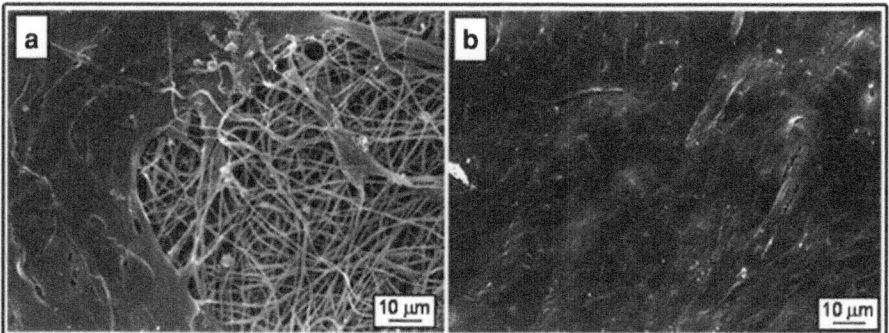

Fig. 3.53 SEM micrographs showing the interaction between H9c2 cells and PPDL ES scaffold after 14 days (**a**) and 27 days of culture (**b**) (Reprinted from [143], Copyright (2010), with permission from Brill Publisher)

differentiation towards endothelial lineage. ES mats supplemented with ECGS where exposed to a culture of Mesenchymal Cells, derived from human bone marrow,[12] with the aim to evaluate the effect of ECGS on cell behavior. The final goal was to assess if ECGS retains its bioactivity once released in the culture medium from the ES fibres [97].

MSCs were cultured[13] in the presence of both plain scaffolds (0-ECGS) and of ECGS-loaded scaffolds (0.4-ECGS and 4.8-ECGS, where the number indicates the ECGS content in wt%). In the course of this experiment cells were not seeded on the ES scaffold but on 2D TCPS, while the mat floated in the culture medium. This procedure allowed to investigate the influence of the GF released from the fibres on cell growth, without introducing any additional effects on cell culture, such as the nature of a 3D substrate. Control experiments were also run without ES mat in the culture dish: as a positive control, cell culture was supplemented with ECGS in a concentration routinely used in cell culture (100 μg/ml); as a negative control an identical experiment was run without ECGS addition. Cell proliferation, cell spreading and shape were evaluated after 7 days of culture.

Cell proliferation was evaluated: (1) through the estimation of the percentage of cells expressing the intranuclear protein Ki-67[14] [145] and (2) by directly counting cell number on images acquired by optical microscopy.[15] Figure 3.54 shows that the percentage of cells expressing the Ki-67 after 7 days of culture was higher in both the positive control and in the 0.4-ECGS and 4.8-ECGS experiments when compared to the percentage of Ki-67 positive cells seen in negative control and 0-ECGS. This result indicates that cells were stimulated to proliferate when cultured in the presence of directly supplemented ECGS, as well as of ECGS-loaded scaffolds.

---

[12] The bone marrow collected from a healthy donor was diluted in culture medium (Medium 199, Sigma–Aldrich) supplemented with 0.5% Fetal Calf Serum (FCS, Sigma–Aldrich) and 1000 U/ml heparin. The diluted bone marrow was layered over a Ficoll-Paque solution and centrifuged at 1100 RPM for 30 min. The cells were washed twice in culture medium, suspended in Medium 199 supplemented with 10% FCS and seeded in 25 cm^2 culture dishes. After 24 h the non adherent cells were removed, while the adherent cells were cultured at 37 °C in 5% CO_2 atmosphere. The culture medium was changed every 2–3 days. After a week a cell monolayer was achieved.

[13] Approximately 1×10^6 cells were seeded in 25 cm^2 TCPS culture dishes in 15 ml of Medium 199 (Sigma–Aldrich) supplemented with 10% Fetal Calf Serum (FCS, from Sigma) for 7 days. The medium was changed every 2–3 days.

[14] Cells were detached from TCPS using 0.25% Trypsin/EDTA solution and the intranuclear Ki-67 protein was estimated by flow cytometry (Beckman Coulter FC 500 Flow Cytometer equipped with argon laser, $\lambda = 488$ nm) by using a Fluorescein FITC conjugated monoclonal antibody (Becton–Dickinson & Co Industry). Values were given as mean percentage cells positive to Ki-67 ± standard error of the mean (sem) and the unpaired t-test was used to evaluate statistical differences between mean values.

[15] Images of the cells were taken at the beginning of the experiments (day 0) and at the end of the 7th day. Cell counting was performed using an AxioObserver microscope (Zeiss) by means of the image analysis software AxioVision 4.6 (Zeiss). Cell number was measured on 350 μm × 350 μm squared areas and the results obtained from 4 different areas per sample were averaged. Values were given as mean ± standard deviation.

Fig. 3.54 Percentage of MSCs cells expressing Ki67 after 7 days of culture (Reprinted from [97], Copyright (2009), with permission from e-polymers)

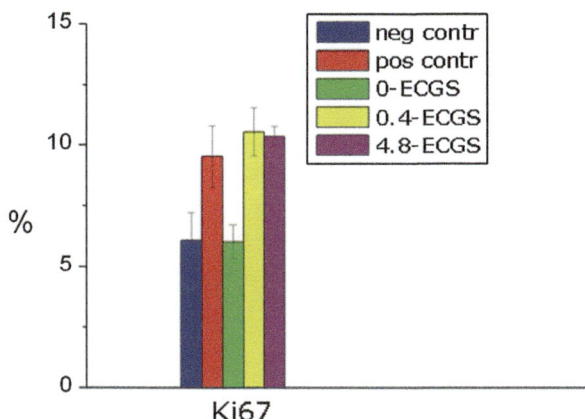

Table 3.6 Cell number changes after 7 days of culture

Experiment	Cell number[a]		Cell number increment (%)
	Day 0	Day 7	
Negative control[b]	79 (9)	86 (4)	+ 9
Positive control[c]	81 (8)	115 (3)	+ 42
0-ECGS[d]	54 (4)	59 (2)	+ 9
0.4-ECGS[d]	67 (2)	97 (4)	+ 45
4.8-ECGS[d]	54 (3)	85 (2)	+ 57

Reprinted from [97], Copyright (2009), with permission from e-polymers
[a] From Differential Interference Contrast Microscopy, standard deviation in parenthesis
[b] Cell culture without ECGS addition
[c] Cell culture with ECGS addition (100 µg/ml)
[d] Cell culture containing the indicated ES fibre mat

Cell proliferation was also estimated by counting the number of cells attached to the TCPS at the beginning of the experiment (day 0) and after 7 days of culture. Cell density varied from sample to sample so cell proliferation was estimated by calculating the percentage increment of cell number. Table 3.6 lists cell number at day 0 and at day 7 as well as the observed percentage increment. It is clear that both negative control and culture containing the scaffold without ECGS (0-ECGS) underwent a cell increment of less than 10%, whereas the cell increment reached around 50% in all other experiments (positive control, 0.4-ECGS, 4.8-ECGS). This result agrees with the trend of the percentage of cells expressing Ki-67 discussed above (Fig. 3.54).

Cell spreading and cell shape, which are known to be strong determinants of the cell fate [146, 147], were investigated by direct observation of cell culture images acquired by DIC microscopy. Figure 3.55 reports some representative images of cell cultured in the presence of ES meshes (0-ECGS, 0.4-ECGS and 4.8-ECGS) together with negative and positive controls. It is clearly observed that the exposure to ECGS (either directly supplemented in the positive control or released from the ECGS-loaded scaffolds) not only promoted cell growth but also affected cell area

Fig. 3.55 Comparison of DIC micrographs at day 0 and day 7, in different culture conditions. Scale bar = 20 μm (Reprinted from [97], Copyright (2009), with permission from e-polymers)

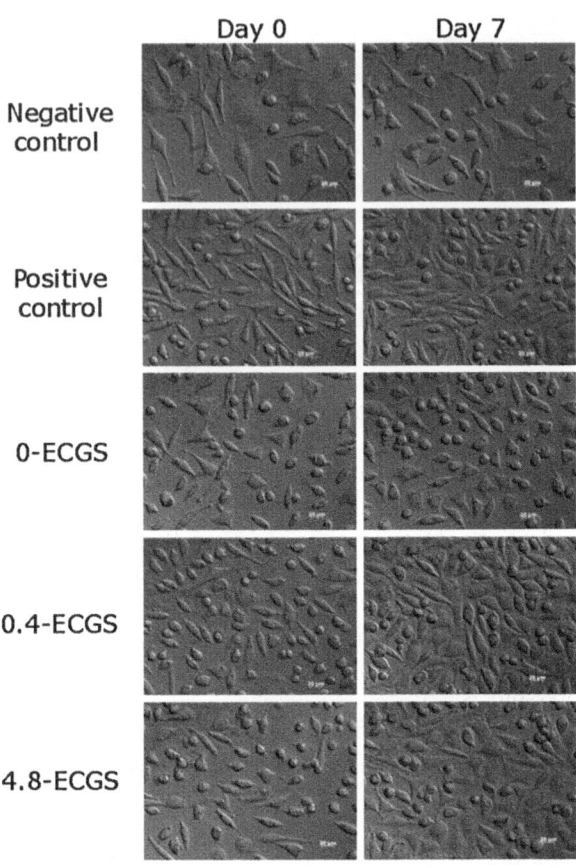

and overall cell shape. By measuring cell area (A) and cell perimeter (P)[16] the cell shape index, that is directly correlated to cell fate, was calculated as follows:

$$S = \frac{4\pi A}{P^2} \tag{3.2}$$

The shape index of an ideal circle is 0, while that of a linear object is 1.

Table 3.7 reports both cell areas and cell shape indexes calculated for all the cell culture conditions at day 0 and at day 7. The shape index of the cells in the negative control and that of the cells cultured in the presence of 0-ECGS slightly increased after 7 days. On the contrary, the shape index of cells from the cultures containing ECGS-loaded mats decreased after 7 days from 0.61 to 0.49 and from 0.74 to 0.41, in 0.4-ECGS and 4.8-ECGS respectively. These values were

[16] Area and perimeter were individually estimated for each cell contained in the four analyzed 350 μm × 350 μm TCPS portions, and the average A and P values were used in Eq. 3.2.

Table 3.7 Cell morphology changes after 7 days of cell culture

Experiment	Shape index[a]		Cell area (μm^2)[b]	
	Day 0	Day 7	Day 0	Day 7
Negative control[c]	0.45 (0.04)	0.64 (0.02)	346 (8)	352 (6)
Positive control[d]	0.67 (0.03)	0.41 (0.03)	342 (3)	416 (5)
0-ECGS[e]	0.56 (0.05)	0.59 (0.02)	343 (3)	366 (3)
0.4-ECGS[e]	0.61 (0.03)	0.49 (0.03)	392 (4)	474 (7)
4.8-ECGS[e]	0.74 (0.03)	0.41 (0.02)	354 (4)	401 (2)

Reprinted from [97], Copyright (2009), with permission from e-polymers
[a] Calculated according to Eq. 3.2, standard deviation in parenthesis
[b] Measured for each cell contained in four analyzed 350 μm × 350 μm cell TCPS portions, standard deviation in parenthesis
[c] Cell culture without ECGS addition
[d] Cell culture with ECGS addition (100 μg/ml)
[e] Cell culture containing the indicated ES fibre mat

comparable to those seen in the positive control where the shape index also decreased from 0.67 (day 0) to 0.41 (day 7). The shape index can assume two limit values, namely 0 and 1, corresponding to circular and linear shape respectively. Hence, the decrease of shape index observed in the ECGS-supplemented culture and in cultures containing ECGS-loaded mats seems to indicate a tendency of the cells to adopt non-elongated shapes.

Visual inspection of the pictures in Fig. 3.55 shows that at day 0 most of the cells had round or elongated shape. At day 7, in the positive control and in the cultures with 0.4-ECGS and 4.8-ECGS a rather clear change in cell shape could be appreciated. Many cells assumed a romboidal shape, in agreement with the mentioned shape index change. In culture, mesenchymal cells can assume distinct yet subtle shape differences in relation to their phenotype lineage; for instance, the shape index of endothelial-like cells is expected to be smaller (in the range 0–0.5) than that of fibroblast-like cells (range 0.5–1). Whether the observed cell shape change could correspond to a true phenotypic shift toward an endothelial-like lineage remains to be established. Taken together these data indicate that ECGS delivered from the scaffolds retains an activity comparable to that of the positive control. ES methodology is therefore suitable for producing bioactive nanofibrous scaffolds through the incorporation of GFs than can be released without loss of bioactivity.

3.4.5 Effect of Electrospun Fibre Orientation on Cancer Cell Culture

In the course of the present Ph.D. ES materials have been designed and fabricated to pursuit the scopes of tissue engineering. This final section illustrates an alternative application of ES materials which, similarly to TE, will probably have a great impact in the future of life sciences. A relatively new application of 3D scaffolds concerns the development of new in vitro models for the study of disease pathogenesis [148, 149]. Indeed, essential cellular functions, that are normally displayed in tissues, are not observable when cells are cultured in TCPS plates (i.e. on a flat 2D surface). Therefore, to date, serious limitations in predicting the cellular responses in the organisms exist. As previously discussed in Chap. 1, when compared to 2D TCPS in vitro cultures, the phenotype assumed by cells on 3D substrates better resembles that developed in vivo. Therefore, the use of 3D scaffolds as new in vitro models is expected to bridge the gap between 2D flat cell culture and living tissues and, consequently, enormous progress in the field of drug screening and reduction in the use of animal models may be foreseen [150].

One of the fields that is currently under investigation is the use of scaffolds as 3D in vitro cell culture models to improve tumour modelling [151]. In vitro cancer models are valuable systems for studying cell behaviour under controlled conditions and under specific therapeutic treatments for circumventing the complexity of in vivo systems. However, researches in cancer pathologies are limited by the experimental instruments currently available for studying these complex diseases. Cancer cells, which are abnormal cells displaying an uncontrolled growth, have the ability to invade surrounding tissues with a large variety of migration modalities. The hallmark of cancer malignity in vivo is invasion. Tumour cells mainly migrate in form of aggregates that assume different shapes [152]. For the sake of simplicity, two main types of cell aggregation can be identified: (1) nest aggregates and (2) chain aggregates. The latter migration mode is commonly associated with tumours exhibiting high level of malignity, because cell penetration into the surrounding tissues is extremely effective. To date it is not clear yet which are the chemical and/or the morphological signals that address cells to a specific migration mode.

With the aim to reproduce an in vitro environment resembling the in vivo ECM organization, ES scaffolds were used as 3D porous substrates to culture mammalian cancer cells. ES mats composed of differently oriented P(L)LA fibres (both randomly arranged and highly aligned) were produced.[17] MCF7 cells derived from a mammary carcinoma cell line[18] were used and cell culture on the ES scaffolds was carried out for 7 days.[19] This experiment aims at investigating the effect of

---

[17] A P(L)LA solution (13% w/V) in DCM/DMF (65:35, by volume) was electrospun using the following process conditions: $\Delta V = 12$ kV, R = 0.015 ml/min, d = 15 cm, T = 22 ± 1 °C, RH = 35 ± 3%. Meshes with randomly arranged fibres were obtained by collecting the fibres on a cylindrical target (radius = 25 mm) rotating at angular rate $\omega = 200$ rpm. Meshes made of highly aligned fibres were obtain by collecting the fibres on the same cylindrical target rotating at angular rate $\omega = 6200$ rpm.

[18] Human breast cancer cell line MCF7 were maintained in RPMI-1640 medium supplemented with 10% fetal bovine serum (FBS), 100 μg/ml Streptomycin, 100 IU/ml Penicillin, 2 mM L-Glutamine and 0.1 mM non-essential amino acids in T-75 flasks in an incubator at 37 °C and 5% CO_2. Sub-confluence cell were trypsinized (0.05% Trysin-EDTA), centrifuged and re-suspended in culture medium.

[19] 1×10^5 breast cancer cells MCF7 were seeded in 1 mL of complete RPMI-1640 medium onto the PLLA ES scaffolds and cultured in standard culture conditions (37 °C, 5% CO_2) for 7 days.

Fig. 3.56 Evaluation of cell adhesion and proliferation on both ES P(L)LA random and aligned fibres compared with polystyrene control by Alamar Blue fluorescence assay

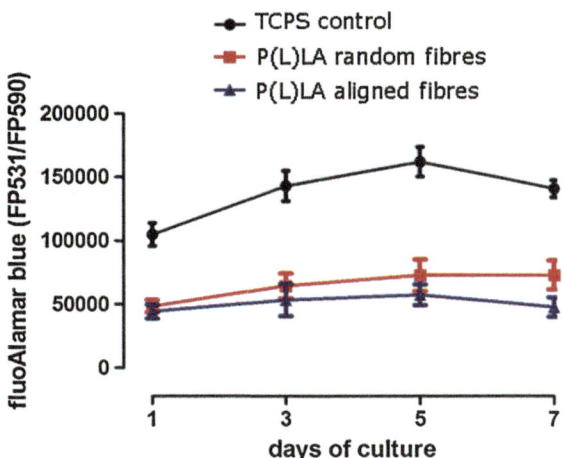

fibre spatial arrangement on cancer cell behaviour in order to better understand the role of the fibrous component of ECM in driving the development of tumour forms in the organism.

Cell adhesion and proliferation on mats composed of differently oriented fibres were evaluated by Alamar Blue assay and results were compared to culture experiment performed on TCPS as a control (Fig. 3.56). The number of viable cells was quantified every other day, up to 7 days. After 24 h from cell seeding, both ES P(L)LA mats (i.e. mat composed of randomly oriented fibres and mat composed of aligned fibres) host about 50% of the number of cells that adhere to the control TCPS surface (day 1). The number of cells growing onto both P(L)LA mats slightly increased in the following days, although a significant difference with respect to the number of the cells proliferating onto the TCPS is maintained at any given time point (p < 0.0001 by ANOVA). Alamar Blue assays showed that P(L)LA fibres are able to host MCF7 cells and that cell proliferation is not affected by scaffold morphology.

Interestingly and contrary to cell proliferation, cell morphology appeared to be highly dependent on the geometry of the substrate. Figure 3.57 shows SEM micrographs of MCF7 after 7 days of culture on P(L)LA fibres differently oriented (both random and aligned fibres) at different magnifications. Cancer cells aggregated on both types of substrates. On random fibres cells generated nest aggregates (Fig. 3.57a, c and e, different magnifications) whereas on aligned fibres both nests (Fig. 3.57d, black circles) and chains (Fig. 3.57d, white circles) could be observed. Higher magnification micrographs clearly shows the different aggregate morphologies assumed by cancer cells on random fibres (Fig. 3.57e) and on aligned fibres (Fig. 3.57f, where a detail of aggregated cells in single row is shown). It is worth noting that, on aligned fibres, both chain aggregates and single cells (white arrows in Fig. 3.57f) are oriented along fibre direction. The tendency of a cell to orient its cytoplasm towards fibre direction when cultured on aligned fibrous substrates has been observed for several cell types [39, 153–156] while the

P(L)LA random fibres   P(L)LA aligned fibres

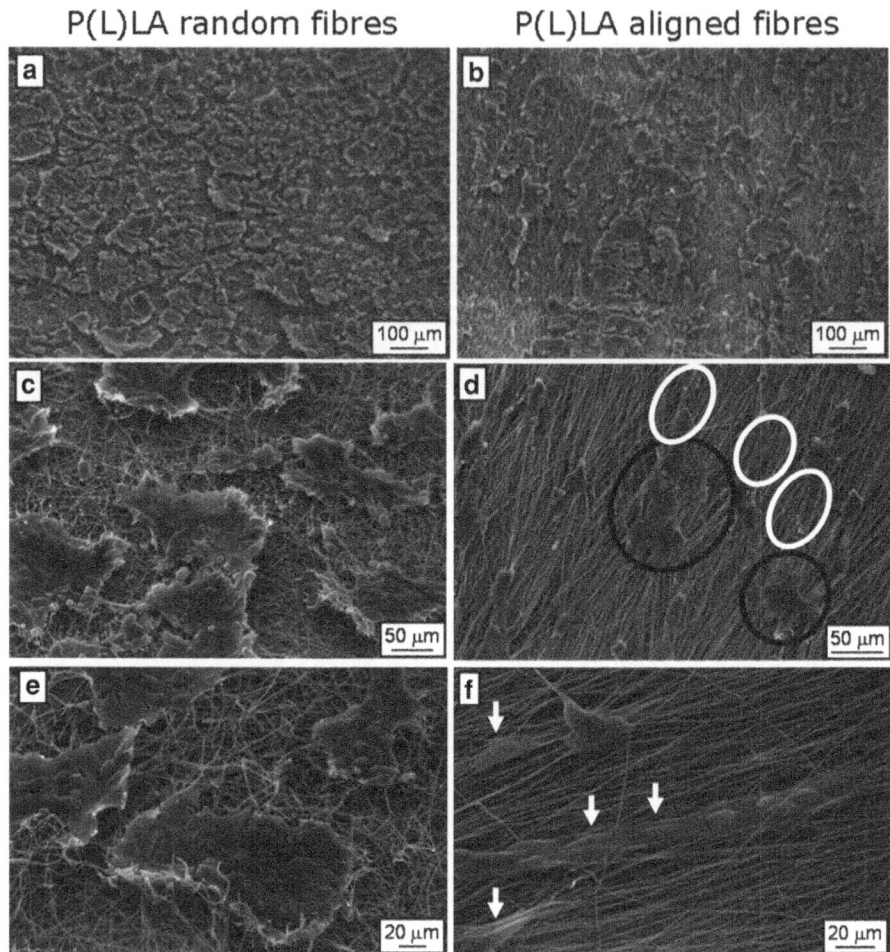

Fig. 3.57 SEM micrographs of MCF7 after 7 days of culture on P(L)LA differently oriented fibres. Cancer cells generated nest aggregate when cultured on random fibres (**a**, **c** and **e**, different magnifications) whereas they generated both nests (*dark circles*) and chains (*white circles*) when cultured on aligned fibres (**b**, **d** and **f**, different magnifications)

cooperative orientation of cells in order to generate an oriented aggregate has not been documented yet.

These preliminary results seem to indicate a strong influence of substrate geometry on cancer cell collective behaviour. Since ES fibrous scaffolds mimic the organization of ECM fibrous component, this finding might be a key information in better understanding the role of natural ECM on tumour malignity in vivo. Moreover, it is pointed out that, if nest aggregates are easily observed on 2D TCPS in vitro cultures, chain aggregates cannot be reproduced in vitro by using the currently available TCPS flat substrates. On the contrary, the use of 3D ES scaffolds helps reproducing the morphologies that MCF7 cells can assume in vivo, demonstrating to be a powerful substrate for modelling cancer cell response in vitro. However, results provided by Alamar Blue assays and SEM observations are far from being exhaustive. A deeper investigation of cancer cell behaviour on 3D ES scaffolds will be carried out in the next future.

References

1. Eichhorn SJ, Sampson WW (2005) Statistical gometry of pores and statistics of porous nanofibrous assemblies. J R Soc Interface 2:309
2. Barry JJA, Gidda HS, Scotchford CA, Howdle SM (2004) Porous methacrylate scaffolds: supercritical fluid fabrication and in vitro chondrocyte responses. Biomaterials 25:3559
3. Barry JJA, Silva MCG, Cartmell SH, Guldberg RE, Scotchford CA, Howdle SM (2006) Porous methacrylate tissue engineering scaffolds: using carbon dioxide to control porosity and interconnectivity. J Mater Sci 41:4197
4. Barry JJA, Nazhat SN, Rose FRAJ, Hainsworth AH, Scotchford CA, Howdle SM (2005) Supercritical carbon dioxide foaming of elastomer/heterocyclic methacrylate blends as scaffolds for tissue engineering. J Mater Chem 15:4881
5. Tai H, Mather ML, Howard D, Wang W, White LJ, Crowe JA, Morgan SP, Williams DJ, Howdle SM, Shakesheff KM (2007) Control of pore size and structure of tissue engineering scaffolds produced by supercritical fluid processing. Eur Cell Mater 14:64
6. Lin ASP, Barrows TH, Cartmell SH, Guldberg RE (2003) Microarchitectural and mechanical characterization of oriented porous polymer scaffolds. Biomaterials 24:481
7. Howdle SM, Watson MS, Whitaker MJ, Popov VK, Davies MC, Mandel FS, Don Wang J, Shakesheff KM (2001) Supercritical fluid mixing: preparation of thermally sensitive polymer composites containing bioactive materials. Chem Commun 1:109
8. Mather ML, Brion M, White LJ, Shakesheff KM, Howdle SM, Morgan SP, Crowe JA (2009) Time-lapsed imaging for in-process evaluation of supercritical fluid processing of tissue engineering scaffolds. Biotechnol Prog 25:1176
9. Singh L, Kumar V, Ratner BD (2004) Generation of porous microcellular 85/15 poly (dl-lactide-co-glycolide) foams for biomedical applications. Biomaterials 25:2611
10. Arora KA, Lesser AJ, McCarthy TJ (1998) Preparation and characterization of microcellular polystyrene foams processed in supercritical carbon dioxide. Macromolecules 31:4614
11. Goel SK, Beckman EJ (1994) Generation of microcellular polymeric foams using supercritical carbon dioxide. I: effect of pressure and temperature on nucleation. Polym Eng Sci 34:1137
12. Mooney DJ, Baldwin DF, Suh NP, Vacanti JP, Langer R (1996) Novel approach to fabricate porous sponges of poly(D, L-lactic-co-glycolic acid) without the use of organic solvents. Biomaterials 17:1417
13. Davies OR, Lewis AL, Whitaker MJ, Tai H, Shakesheff KM, Howdle SM (2008) Applications of supercritical CO_2 in the fabrication of polymer systems for drug delivery and tissue engineering. Adv Drug Deliv Rev 60:373
14. Oliveira NS, Dorgan J, Coutinho JAP, Ferreira A, Daridon JL, Marrucho IM (2006) Gas solubility of carbon dioxide in poly(lactic acid) at high pressures. J Polym Sci Part B Polym Phys 44:1010
15. Gualandi C, White LJ, Chen L, Gross RA, Shakesheff KM, Howdle SM, Scandola M (2010) Scaffold for tissue engineering fabricated by non-isothermal supercritical carbon dioxide foaming of a highly crystalline polyester. Acta Biomater 6:130
16. Ceccorulli G, Scandola M, Kumar A, Kalra B, Gross RA (2005) Cocrystallization of random copolymers of ω-pentadecalactone and ε-caprolactone synthesized by lipase catalysis. Biomacromolecules 6:902

17. Doroudiani S, Park CB, Kortschot MT (1996) Effect of the crystallinity and morphology on the microcellular foam structure of semicrystalline polymers. Polym Eng Sci 36:2645
18. Tomasko DL, Li H, Liu D, Han X, Wingert MJ, Lee LJ, Koelling KW (2003) A review of CO_2 applications in the processing of polymers. Ind Eng Chem Res 42:6431
19. Mathieu L, Montjovent M-O, Bourban P-E, Pioletti DP, Manson J-AE (2005) Bioresorbable composites prepared by supercritical fluid foaming. J Biomed Mater Res Part A 75:89
20. Fox TG (1956) Influence of diluent and of copolymer composition on the glass temperature of a polymer system. Bull Am Phys Soc 1:123
21. Couchman PR, Karasz FE (1978) A classical thermodynamic discussion of the effect of composition on glass-transition temperatures. Macromolecules 11:117
22. Chow TS (1980) Molecular interpretation of the glass transition temperature of polymer-diluent systems. Macromolecules 13:362
23. Flory PJ (1953) Principles of polymer chemistry. Cornell University Press, New York
24. Gibson LJ (2005) Biomechanics of cellular solids. J Biomech 38:377
25. Black J, Hastings G (1998) Handbook of biomaterial properties. Chapman & hall, London
26. Kutz M (2003) Standard handbook of biomedical engineering and design. The McGraw-Hall Companies, New York
27. Fridrikh SV, Yu JH, Brenner MP, Rutledge G (2003) Controlling the fiber diameter during electrospinning. Phys Rev Lett 90:144502-1
28. Shin M, Hohman MM, Brenner MP, Rutledge G (2001) Electrospinning: a whipping fluid jet generates submicron polymer fibers. Appl Phys Lett 78:1149
29. Helgeson ME, Grammatikos KN, Deitzel JM, Wagner NJ (2008) Theory and kinematic measurement of the mechanics of stable electrospun polymer jets. Polymer 49:2924
30. Reneker DH, Yarin AL, Fong H, Koombhongse S (2000) Bending instability of electricaly charged liquid jets of polymer solutions in electrospinning. J Appl Phys 87:4531
31. Shin YM, Hohman MM, Brenner MP, Rutledge GC (2001) Experimental characterization of electrospinning: the electrically forced jet and instabilities. Polymer 42:9955
32. Reneker DH, Yarin AL, Zussman E, Koombhongse S, Kataphinan W (2006) Nanofiber manufacturing: toward better process control. In: Reneker DH, Fong H (eds) Polymeric Nanofibers, ACS Symposium Series 918. Oxford University Press, Oxford, pp 7–20
33. Taylor G (1969) Electrically driven jets. Proc R Soc Lond 313:453
34. Lee JS, Choi KH, Ghim HD, Kim S-S, Chun DH, Kim HY, Ma K (2004) Role of molecular weight of atactic poly(vinyl alcohol) (PVA) in the structure and properties of PVA nanofabric prepared by electrospinning. J Appl Polym Sci 93:1638
35. Tan S-H, Inai R, Kotaki M, Ramakrishna S (2005) Systematic parameter study for ultra-fine fiber fabrication via electrospinning process. Polymer 46:6128
36. Katti DS, Robinson KW, Ko FK, Laurencin CT (2004) Bioresorbable nanofiber-based systems for wound healing and drug delivery: optimization of fabrication parameters. J Biomed Mater Res B Appl Biomater 70B:286
37. Megelski S, Stephens JS, Chase DB, Rabolt JF (2002) Micro- and nanostructured surface morphology on electrospun polymer fibers. Macromolecules 35:8456
38. Kidoaki S, Kwon IK, Matsuda T (2006) Structural features and mechanical properties of in situ–bonded meshes of segmented polyurethane electrospun from mixed solvents. J Biomed Mater Res B Appl Biomater 76B:219
39. Baker SC, Atkin N, Gunning PA, Granville N, Wilson K, Wilson D, Southgate J (2006) Characterisation of electrospun polystyrene scaffolds for three-dimensional in vitro biological studies. Biomaterials 27:3136
40. Gupta P, Elkins C, Long TE, Wilkes GL (2005) Electrospinning of linear homopolymers of poly(methyl methacrylate): exploring relationships between fiber formation, viscosity, molecular weight and concentration in a good solvent. Polymer 46:4799
41. Shenoy SL, Bates WD, Frisch HL, Wnek GE (2005) Role of chain entanglements on fiber formation during electrospinning of polymer solutions: good solvent, non-specific polymer-polymer interaction limit. Polymer 46:3372

42. Jarusuwannapoom T, Hongrojjanawiwat W, Jitjaicham S, Wannatong L, Nithitanakul M, Pattamaprom C, Koombhongse P, Rangkupan R, Supaphol P (2005) Effect of solvents on electro-spinnability of polystyrene solutions and morphological appearence of resulting electrospun polystyrene fibers. Eur Polym J 41:409
43. Son WK, Youk JH, Lee TS, Park WH (2004) The effects of solution properties and polyelectrolyte on electrospinning of ultrafine poly(ethylene oxide) fibers. Polymer 45:2959
44. Lide DR (2007) Handbook of chemistry and physics. CRC Press, Boca Raton
45. Zhong XH, Kim K, Fang D, Ran S, Hsiao BS, Chu B (2002) Structure and process relationship of electrospun bioabsorbable nanofiber membranes. Polymer 43:4403
46. Fong H, Reneker DH (1999) Beaded nanofibers formed during electrospinning. Polymer 40:4585
47. Ramakrishna S, Fujihara K, Teo W-E, Lim T, Ma Z (2005) An introduction to electrospinning and nanofibers. World Scientific Publishing, Singapore
48. Lee KH, Kim HY, Bang HJ, Jung YH, Lee SG (2003) The change of bead morphology formed on electrospun polystyrene fibers. Polymer 44:4029
49. Vert M, Schwarch G, Czerwonka LA (1995) Present and future of PLA polymers. J Macromol Sci Chem 32:787
50. Middleton JC, Tipton AJ (2000) Synthetic biodegradable polymers as orthopedic devices. Biomaterials 21:2335
51. Engelberg I, Kohn J (1991) Physico-mechanical properties of degradable polymers used in medical applications: a comparative study. Biomaterials 12:292
52. Fischer EW, Sterzel HJ, Wegner G (1973) Investigation of the structure of solution grown crystals of lactide copolymers by means of chemical reactions. Kolloid-Z u Z Polymere 251:980
53. Garlotta D (2001) A literature review of poly(lactic acid). J Environ Polym Degr 9:63
54. Henton DE, Gruber PR, Lunt J, Randall J (2005) Polylactic acid technology. In: Mohanty AK, Misra M, Drzal LT (eds) Natural Fibers, Biopolymers and Biocomposites. CRC Press, Boca Raton, pp 527–578
55. Pego AP, Poot AA, Grijpma DW, Feijen J (2003) Biodegradable elastomeric scaffolds for soft tissue engineering. J Controlled Release 87:69
56. Plikk P, Malberg S, Albertsson A-C (2009) Design of resorbable porous tubular copolyester scaffolds for use in nerve regeneration. Biomacromolecules 10:1259
57. Geminiani G (2009) Studio di materiali polimerici nanofibrosi con proprietà elastomeriche per usi biomedicali. Chemistry Dept "G. Ciamician", University of Bologna, Chemistry Eurobachelor
58. Pitt CG (1990) Poly-ε-caprolactone and its copolymers. In: Chasin M, Langer R (eds) Biodegradable Polymers as Drug Delivery Systems. Marcel Dekker, New York, pp 71–119
59. Zhong X, Ran S, Fang D, Hsiao BS, Chu B (2003) Control of structure, morphology and property in electrospun poly(glycolide-co-lactide) non-woven membranes via post-draw treatments. Polymer 44:4959
60. Deitzel JM, Kleinmeyer JD, Hirvonen JK, Beck Tan NC (2001) Controlled deposition of electrospun poly(ethylene oxide) fibers. Polymer 42:8163
61. Pitt CG, Chasalow FI, Hibionada YM, Klimas DM (1981) Aliphatic polyesters. I. The degradation of poly(ε-caprolactone) in vivo. J Appl Polym Sci 26:3779
62. Focarete ML, Gazzano M, Scandola M, Gross RA (2002) Copolymers of ω-pentadecalactone and trimethylene carbonate from lipase catalysis: influence of microstructure on solid-state properties. Macromolecules 35:8066
63. Jiang Z, Azim H, Gross RA, Focarete ML, Scandola M (2007) Lipase-catalyzed copolymerization of ω-pentadecalactone with p-dioxanone and characterization of copolymer thermal and crystalline properties. Biomacromolecules 8:2262
64. Shawon J, Sung C (2004) Electrospinning of polycarbonate nanofibers with solvent mixtures THF and DMF. J Mater Sci 39:4605

65. Casper CL, Stephens JS, Tassi NG, Chase DB, Rabolt JF (2004) Controlling surface morphology of electrospun polystyrene fibers: effect of humidity and molecular weight in electrospinning process. Macromolecules 37:573
66. Teo W-E, Ramakrishna S (2006) A review on electrospinning design and nanofibre assemblies. Nanotechnology 17:R89
67. Matthews JA, Wnek GE, Simpson DG, Bowlin GL (2002) Electrospinning of collagen nanofibers. Biomacromolecules 3:232
68. Katta P, Alessandro M, Ramsier RD, Chase GG (2004) Continuous electrospinning of aligned polymer nanofibers onto a wire drum collector. Nano Lett 4:2215
69. Wu Y, Carnell LA, Clark RL (2007) Control of electrospun mat width through the use of parallel auxiliary electrodes. Polymer 48:5653
70. Carnell LS, Siochi EJ, Holloway NM, Stephens RM, Rhim C, Niklason LE, Clark RL (2008) Aligned mats from electrospun single fibers. Macromolecules 41:5345
71. Teo W-E, Kotaki M, Mo XM, Ramakrishna S (2005) Porous tubular structures with controlled fibre orientation using a modified electrospinning method. Nanotechnology 16:918
72. Li D, Wang YL, Xia YN (2004) Electrospinning nanofibers as uniaxially aligned arrays and layer-by-layer stacked films. Adv Mat 16:361
73. Liu L, Dzenis Y (2008) Analysis of the effects of the residual charge and gap size on electrospun nanofiber alignment in a gap method. Nanotechnology 19:355307
74. Li D, Ouyang G, McCann JT, Xia Y (2005) Collecting electrospun nanofibers with patterned electrodes. Nano Lett 5:913
75. Zhang D, Chang J (2007) Patterning of electrospun fibers using electroconductive templates. Adv Mater 19:3664
76. Gibson P, Schreuder-Gibson H (2004) Patterned electrospray fiber structures. Int Nonwovens J 13:34
77. Ramakrishna S, Fujihara K, Teo W-E, Lim T, Ma Z (2005) Electrospinning process. In: Ramakrishna S, Fujihara K, Teo W-E, Lim T, Ma Z (eds) An Introduction to Electrospinning and Nanofibers. World Scientific Publishing, Singapore, pp 135–137
78. Zucchelli A, Fabiani D, Gualandi C, Focarete ML (2009) An innovative and versatile approach to design highly porous, micropatterned, nanofibrous polymeric materials. J Mater Sci 44:4969
79. Vargin VV (1967) Technology of enamels. MacLaren and Sons Ltd, London
80. Tavgen VV (2000) Glass ceramic enamels with increased conductivity. Glass Ceram 57:183
81. Kidoaki S, Kwon IK, Matsuda T (2005) Mesoscopic spatial designs of nano- and microfiber meshes for tissue-engineering matrix and scaffold based on newly devised multilayering and mixing electrospinning techniques. Biomaterials 26:37
82. Sala F (2009) Studio delle proprietà meccaniche di supporti polimerici elettrofilati e sviluppo di scaffold tubolari per la rigenerazione del tessuto nervoso. Chemistry Dept "G. Ciamician", University of Bologna, Chemistry Eurobachelor
83. Hong Y, Fujimoto K, Hashizume R, Guan J, Stankus JJ, Tobita K, Wagner WR (2008) Generating elastic, biodegradable polyurethane/poly(lactide-co-glycolide) fibrous sheets with controlled antibiotic release via two-stream electrospinning. Biomacromolecules 9:1200
84. Huang Z-M, He CL, Yang A, Zhang Y, Han X-J, Yin J (2006) Encapsulating drugs in biodegradable ultrafine fibers through co-axial electrospinning. J Biomed Mater Res Part A 77A:169
85. Kim K, Luu YK, Chang C, Fang D, Hsiao BS, Chu B, Hadjiargyrou M (2004) Incorporation and controlled release of a hydrophilic antibiotic using poly(lactide-co-glycolide)-based electrospun nanofibrous scaffolds. J Controlled Release 98:47
86. Kenawy ER, Abdel-Hay FI, El-Newehy MH, Wnek GE (2009) Processing of polymer nanofibers through electrospinning as drug delivery systems. Mater Chem Phys 113:296

87. Piras AM, Nikkola L, Chiellini F, Ashammakhi N, Chiellini E (2006) Development of diclofenac sodium releasing bio-erodible polymeric nanomats. J Nanosci Nanotechnol 6:3310

88. Taepaiboon P, Rungsardthong U, Supaphol P (2006) Drug-loaded electrospun mats of poly(vinyl alcohol) fibres and their release characteristics of four model drugs. Nanotechnology 17:2317

89. Xie J, Wang C-H (2006) Electrospun micro- and nanofibers for sustained delivery of paclitaxel to treat C6 glioma in vitro. Pharm Res 23:1817

90. Casper CL, Yang WD, Farach-Carson MC, Rabolt JF (2007) Coating electrospun collagen and gelatin fibers with perlecan domain I for increased growth factor binding. Biomacromolecules 8:1116

91. Lu Y, Jiang HL, Tu KH, Wang LQ (2009) Mild immobilization of diverse macromolecular bioactive agents onto multifunctional fibrous membranes prepared by coaxial electrospinning. Acta Biomater 5:1562

92. Choi JS, Leong KW, Yoo HS (2008) In vivo wound healing of diabetic ulcers using electrospun nanofibers immobilized with human epidermal growth factor (EGF). Biomaterials 29:587

93. Chew SY, Mi R, Hoke A, Leong KW (2007) Aligned protein-polymer composite fibers enhance nerve regeneration: a potential tissue-engineering platform. Adv Funct Mater 17:1288

94. Chew SY, Wen J, Yim EKF, Leong KW (2005) Sustained release of proteins from electrospun biodegradable fibers. Biomacromolecules 6:2017

95. Su Y, Li XQ, Tan LJ, Huang C, Mo XM (2009) Poly(L-lactide-co-ε-caprolactone) electrospun nanofibers for encapsulating and sustained releasing proteins. Polymer 50:4212

96. Liao IC, Chew SY, Leong KW (2006) Aligned core-shell nanofibers delivering bioactive proteins. Nanomedicine 1:465

97. Gualandi C, Wilczek P, Focarete ML, Pasquinelli G, Kawalec M, Scandola M (2009) Bioresorbable electrospun nanofibrous scaffolds loaded with bioactive molecules. e-Polymers 60

98. Yoo HS, Kim TG, Park TG (2009) Surface-functionalized electrospun nanofibers for tissue engineering and drug delivery. Adv Drug Deliv Rev 61:1033

99. Zhu X, Chian KS, Chan-Park MBE, Lee ST (2005) Effect of argon-plasma treatment on proliferation of human-skin-derived fibroblast on chitosan membrane in vitro. J Biomed Mater Res Part A 73A:264

100. Venugopal J, Low S, Choon AT, Kumar AB, Ramakrishna S (2008) Electrospun-modified nanofibrous scaffolds for the mineralization of osteoblast cells. J Biomed Mater Res Part A 85A:408

101. Prabhakaran MP, Venugopal J, Chan CK, Ramakrishna S (2008) Surface modified electrospun nanofibrous scaffolds for nerve tissue engineering. Nanotechnology 19:455102

102. He W, Ma ZW, Yong T, Teo WE, Ramakrishna S (2005) Fabrication of collagen-coated biodegradable polymer nanofiber mesh and its potential for endothelial cells growth. Biomaterials 26:7606

103. Koh HS, Yong T, Chan CK, Ramakrishna S (2008) Enhancement of neurite outgrowth using nano-structured scaffolds coupled with laminin. Biomaterials 29:3574

104. Choi WS, Bae JW, Lim HR, Joung YK, Park JC, Kwon IK, Park KD (2008) RGD peptide-immobilized electrospun matrix of polyurethane for enhanced endothelial cell affinity. Biomed Mater 3:44104

105. Kim TG, Park TG (2006) Biomimicking extracellular matrix: cell adhesive RGD peptide modified electrospun poly(D, L-lactic-co-glycolic acid) nanofiber mesh. Tissue Eng 12:221

106. Park K, Ju YM, Son JS, Ahn KD, Han DK (2007) Surface modification of biodegradable electrospun nanofiber scaffolds and their interaction with fibroblasts. J Biomater Science-Polymer Edition 18:369

107. Yao C, Li XS, Neoh KG, Shi ZL, Kang ET (2008) Surface modification and antibacterial activity of electrospun polyurethane fibrous membranes with quaternary ammonium moieties. J Membr Sci 320:259
108. Ratner BD, Hoffman AS (2004) Nonfouling surfaces. In: Ratner BD, Hoffman AS, Schoen FJ, Lemons JE (eds) Biomaterials Science: An Introduction to Materials in Medicine. Elsevier Academic press, San Diego, pp 197–200
109. Patrucco E, Ouasti S, Vo CD, De Leonardis P, Pollicino A, Armes SP, Scandola M, Tirelli N (2009) Surface-initiated ATRP modification of tissue culture substrates: poly(glycerol monomethacrylate) as an antifouling surface. Biomacromolecules 10:3130
110. Patten TE, Matyjaszewski K (1998) Atom transfer radical polymerization and the synthesis of polymeric materials. Adv Mater 10:901
111. Matyjaszewski K, Xia J (2001) Atom transfer radical polymerization. Chem Rev 101:2921
112. Xu FG, Neoh KG, Kang ET (2009) Bioactive surfaces and biomaterials via atom transfer radical polymerization. Prog Polym Sci 34:719
113. Fu GD, Lei JY, Yao C, Li XS, Yao F (2008) Core-sheat nanofibers from combined atom transfer radical polymerization and electrospinning. Macromolecules 41:6854
114. Sun X-Y, Shankar R, Borner HG, Ghosh TK, Spontak RJ (2007) Field-driven biofunctionalization of polymer fiber surfaces during electrospinning. Adv Mater 19:87
115. Li D, Frey MW, Vynias D, Baeumner AJ (2007) Availability of biotin incorporated in electrospun PLA fibers for streptavidin binding. Polymer 48:6340
116. Kim SJ, Lee CK, Kim SI (2005) Effect of ionic salts on the processing of poly(2acrylamido-2-methyl-1-propane sulfortic acid) nanofibers. J Appl Polym Sci 96:1388
117. Haigh R, Rimmer S, Fullwood NJ (2000) Synthesis and properties of amphiphilic networks. 1: the effect of hydration and polymer composition on the adhesion of immunoglobulin-G to poly(laurylmethacrylate-stat-glycerolmonomethacrylate-stat-ethylene-glycol-dimethacrylate) networks. Biomaterials 21:735
118. Mequanint K, Patel A, Bezuidenhout D (2006) Synthesis, swelling behavior and biocompatibility of novel physically cross-linked polyurethane-block-poly(glycerol methacrylate) hydrogels. Biomacromolecules 7:883
119. Burkersroda F, Schedl L, Gopferich A (2002) Why degradable polymers undergo surface erosion or bulk erosion. Biomaterials 23:4221
120. Li S (1999) Hydrolytic degradation characteristics of aliphatic polyesters derived from lactic and glycolic acid. J Biomed Mater Res 48:342
121. Ikada Y, Jamshidi K, Tsuji H, Hyon SH (1987) Stereocomplex formation between enantiomeric poly(lactides). Macromolecules 20:904
122. Li S, Garreau H, Vert M (1990) Structure-property relationships in the case of the degradation of massive aliphatic poly-(α-hydroxy acids) in aqueous media, part 2:degradation of lactide-glycolide copolymers: PLA37.5GA25 and PLA75GA25. J Mater Sci Mater Med 1:131
123. Li S, Garreau H, Vert M (1990) Structure-property relationships in the case of the degradation of massive aliphatic poly-(α-hydroxy acids) in aqueous media, part 3: influence of the morphology of poly(L-lactic acid). J Mater Sci Mater Med 1:198
124. Li S, Garreau H, Vert M (1990) Structure-property relationships in the case of the degradation of massive aliphatic poly-(α-hydroxy acids) in aqueous media, part 1: poly (D, L-lactic acid). J Mater Sci Mater Med 1:123
125. Den Dunnen WFA, Robinson PH, van Wessel R, Pennings AJ, van Leeuwen M, Schakenraad JM (1997) Long-term evaluation of degradation and foreign-body reaction of subcutaneously implanted poly(DL-lactide-ε-caprolactone). J Biomed Mater Res 36:337
126. Vert M, Li SM, Spenlehauer G, Guerin P (1992) Bioresorbability and biocompatibility of aliphatic polyesters. J Mater Sci Mater Med 3:432
127. Li WJ, Cooper J, Mauck RL, Tuan RS (2006) Fabrication and characterization of six electrospun poly(α-hydroxy ester)-based fibrous scaffolds for tissue engineering applications. Acta Biomater 2:377

128. Zong X, Ran S, Kim K-S, Fang D, Hsiao BS, Chu B (2003) Structure and morphology changes during in vitro degradation of electrospun poly(glycolide-co-lactide) nanofiber membrane. Biomacromolecules 4:416
129. Xu CY, Inai R, Kotaki M, Ramakrishna S (2004) Aligned biodegradable nanofibrous structure: a potential scaffold for blood vessel engineering. Biomaterials 25:877
130. Zhang XH, Baughman CB, Kaplan DL (2008) In vitro evaluation of electrospun silk fibroin scaffolds for vascular cell growth. Biomaterials 29:2217
131. Li M, Guo Y, Wei Y, MacDiarmid AG, Lelkes PI (2006) Electrospinning polyaniline-contained gelatin nanofibers for tissue engineering applications. Biomaterials 27:2705
132. Zhang J, Qi HX, Wang HJ, Hu P, Ou LL, Guo SH, Li J, Che YZ, Yu YT, Kong DL (2006) Engineering of vascular grafts with genetically modified bone marrow mesenchymal stem cells on poly (propylene carbonate) graft. Artif Organs 30:898
133. Chiu JB, Liu C, Hsiao BS, Chu B, Hadjiargyrou M (2007) Functionalization of poly(L-lactide) nanofibrous scaffolds with bioactive collagen molecules. J Biomed Mater Res 83A:1117
134. Williamson MR, Black RA, Kielty C (2006) PCL-PU composite vascular scaffold production for vascular tissue enginerring: attachment, proliferation and bioactivity of human vascular endothelial cells. Biomaterials 27:3608
135. Zeng J, Xu X, Chen X, Liang Q, Bian X, Yang L, Jing X (2003) Biodegradable electrospun fibers for drug delivery. J Controlled Release 92:227
136. Loebsack AB, Halberstadt CR, Holder WD, Culberson CR, Beiler RJ, Greene KG, Roland WD, Burg KJL (1999) The development of an embedding technique for polylactide sponges. J Biomed Mater Res 48:504
137. Brown DA, Chou YF, Beygui RE, Dunn JCY, Wu BM (2005) Gelatin-embedded cell-polymer constructs for histological cryosectioning. J Biomed Mater Res Part B-Appl Biomater 72B:79
138. Holy CE, Yakubovich R (2000) Processing cell-seeded polyester scaffolds for histology. J Biomed Mater Res 50:276
139. Baker BM, Mauck RL (2007) The effect of nanofiber alignment on the maturation of engineered meniscus constructs. Biomaterials 28:1967
140. Pham QP, Sharma U, Mikos AG (2006) Electrospun poly(ε-caprolactone) microfiber and multilayer nanofiber/microfiber scaffolds: characterization of scaffolds and measurement of cellular infiltration. Biomacromolecules 7:2796
141. Corey JM, Gertz CC, Wang B-S, Birrell LK, Johnson SL, Martin DC, Feldman EL (2008) The design of electrospun PLLA nanofiber scaffolds compatible with serum-free growth of primary motor and sensory neurons. Acta Biomater 4:863
142. Foroni L, Dirani G, Gualandi C, Focarete ML, Pasquinelli G (2010) Paraffin embedding allows effective analysis of proliferation, survival and immunophenotyping of cells cultured on poly(L-lactic acid) electrospun nanofiber scaffolds. Tissue Eng Part C Methods 16(4):751
143. Focarete ML, Gualandi C, Scandola M, Govoni M, Giordano ED, Foroni L, Valente S, Pasquinelli G, Gao W, Gross RA (2010) Electrospun scaffolds of a polyhydroxyalkanoate consisting of ω-hydroxylpentadecanoate repeat units: fabrication and in vitro biocompatibility studies. J Biomater Sci Polym Edition 21:1283
144. Van der Meulen I, de Geus M, Antheunis H, Deumens R, Joosten EAJ, Koning CE, Heise A (2008) Polymers from Functional Macrolactones as Potential Biomaterials: Enzymatic Ring Opening Polymerization, Biodegradation, Biocompatibility. Biomacromolecules 9:3404
145. Gerdes J, Lemke H, Baisch H, Wacker HH, Schwab U, Stein H (1984) Cell cycle analysis of a cell proliferation-associated human nuclear antigen defined by the monoclonal antibody Ki-67. J Immunol 133:1710
146. Chen CS, Mrksich M, Huang S, Whitesides GM, Ingber DE (1997) Geometric control of cell live and death. Science 276:1425
147. Huang S, Ingber DE (2000) Shape-dependent control of cell growth, differentiation and apoptosis: switching between attractors in cell regulatory networks. Exp Cell Res 261:91

148. Griffith LG, Swartz MA (2006) Capturing complex 3D tissue physiology in vitro. Nat Rev Mol Cell Biol 7:211
149. Hutmacher DW, Horch RE, Loessner D, Rizzi S, Sieh S, Reichert JC, Clements JA, Beler JP, Arkudas A, Bleiziffer O, Kneser U (2009) Translating tissue engineering technology platforms into cancer research. J Cell Mol Med 13:1417
150. Pampaloni F, Reynaud EG, Stelzer EHK (2007) The third dimension bridges the gap between cell culture and live tissue. Nat Rev Mol Cell Biol 8:839
151. Yamada KM, Cukierman E (2007) Modelling tissue morphogenesis and cancer in 3D. Cell 130:601
152. Friedl P, Wolf K (2003) Tumour-cell invasion and migration: diversity and escape mechanisms. Nat Rev Cancer 3:362
153. Rockwood DN, Akins RE, Parrag IC, Woodhouse KA, Rabolt JF (2008) Culture on electrospun polyurethane scaffolds decreases atrial natriuretic peptide expression by cardiomyocytes in vitro. Biomaterials 29:4783
154. Schnell E, Klinkhammer K, Balzer S, Brook G, Klee D, Dalton P, Mey J (2007) Guidance of glial cell migration and axonal growth on electrospun nanofibers of poly-ε-caprolactone and a collagen/poly-ε-caprolactone blend. Biomaterials 28:3012
155. Xie J, Willerth SM, Li X, Macewan MR, Rader A, Sakiyama-Elbert SE, Xia Y (2009) The differentiation of embryonic stem cells seeded on electrospun nanofibers into neural lineages. Biomaterials 30:354
156. Lee CH, Shin HJ, Cho IH, Kang Y-M, Kim IA, Park K-D, Shin J-W (2005) Nanofiber alignment and direction of mechanical strain affect the ECM production of human ACL fibroblast. Biomaterials 26:1261

Chapter 4
Conclusions

Tissue engineering (TE) is a rapidly growing discipline which integrates the basic principles of biology, engineering and material science with the aim to repair or regenerate damaged tissues. To this scope, a cell-construct is engineered in vitro by using a porous 3D material as cell culture support, commonly referred to as scaffold. The role of the scaffold is to act as a temporary template, guiding cell organization, growth and differentiation and providing a structural stability and a 3D environment where cells can produce new biological tissue. Therefore, the scaffold must be designed to be bioresorbed in the organism and replaced by new tissue produced by cells. The obtainment of a successful engineered tissue encompasses the optimization of several critical elements (e.g. cell type, scaffold material and 3D structure, cell culture conditions, etc.) and an effective multidisciplinary approach involving not only biological and medical expertises but also competences in engineering, chemistry and materials science is imperative in this context.

The present Ph.D. project focused its attention on the design, fabrication, manipulation and characterization of innovative polymeric bioresorbable scaffolds made of hydrolysable polyesters. Two techniques were employed to fabricate scaffolds: supercritical carbon dioxide ($scCO_2$) foaming and electrospinning (ES). Microporous 3D foamed materials, that can assume a structural supporting role when used as tissue replacements, will be particularly suitable in hard TE (e.g. bone or cartilage TE), whereas flexible nanofibrous electrospun mats will be more appropriate for replacement of soft tissues such as cardiac, vascular or nervous tissues.

$ScCO_2$ foaming is a powerful scaffold fabrication technique since it does not require the use of solvents that can be toxic to mammalian cells. However, applicability of this technology is mainly limited to amorphous polymers which can be easily processed into foams. In the course of the present Ph.D., it was demonstrated that a proper modification of the $scCO_2$ foaming process usually applied to amorphous materials enables to successfully produce porous structures

C. Gualandi, *Porous Polymeric Bioresorbable Scaffolds for Tissue Engineering*,   119
Springer Theses, DOI: 10.1007/978-3-642-19272-2_4,
© Springer-Verlag Berlin Heidelberg 2011

also from highly crystalline polymers. The investigation of the effect of process parameters on scaffold morphology by Micro X-ray Computed Tomography contributed to demonstrate the versatility of this technique and to broaden the range of polymeric materials that can be processed into foams using $scCO_2$ as porogen (Sect. 3.1).

The present research activity mainly focused onto the study and the development of ES process and of electrospun products. The main benefit in using ES technique is its outstanding versatility: it is documented that electrospun fibres from over two hundred synthetic and natural polymers can be produced. Another advantage is the simplicity of the instrumental apparatus that does not require any sophisticated and expensive equipment. Moreover, given the intrinsic morphologically biomimetic features of electrospun materials, which are composed of sub-micrometric fibres resembling the fibrous component of native tissue, they are considered the most promising scaffolds by most of the scientific community. In the course of the present Ph.D., a highly reproducible ES process, performed in a glove box specifically designed to control the environmental conditions, was developed for fabricating innovative electrospun scaffolds. New non-commercial polymers were electrospun for the first time. These polymers, being more hydrophobic than common commercial bioresorbable polyesters, might be particularly interesting for long-term applications. Innovative patterned electrospun non-woven meshes were also produced by using novel ad hoc designed ES collectors that enabled a fine control of fibre deposition geometry. Particularly interesting is the possibility to obtain different fibre patterns contemporarily available in a given scaffold. Such type of scaffold is suitable for further investigation concerning the effect of substrate morphology on cell homing, spreading and organization.

Besides the improvement of ES process, part of the research activity concerned the production of bio-functionalized electrospun scaffolds that was achieved by either a bulk functionalization or a surface functionalization. A bulk functionalization was successfully performed by incorporating a growth factor (GF) within electrospun fibres demonstrating that ES technology is suitable for producing bioactive nanofibrous scaffolds containing GFs, which can be subsequently released without loss of bioactivity. The surface modification approach developed in the course of the present research activity aimed at controlling cell adhesion on electrospun scaffolds. The modification was achieved by covering fibre surface with highly hydrophilic polymer brushes carrying hydroxyl groups. The hydrophilic nature of the brushes is expected to passivate fibre surface towards protein absorption, and thus towards cell attachment. The preliminary results presented in this thesis open a wide range of possibility to exploit the hydroxyl groups of the brushes for functionalizing fibre surface with biomolecules, such as peptides or GFs, for eliciting the desired cell behaviour.

Collaboration with biochemical laboratories allowed to perform cell culture experiments by using some of the electrospun scaffolds produced in the course of this Ph.D. Finally, the last part of the research activity was dedicated to explore the possibility to use electrospun materials in an application alternative to TE that will probably have a great impact in the future of life sciences. Indeed, very recently, 3D scaffolds have been employed for the development of new in vitro tumour models. Preliminary cell culture experiments carried out in the course of this research demonstrated a strong influence of fibre orientation on cancer cell collective behaviour. This finding might provide a key information for better understanding the role of natural ECM on tumour malignity in vivo. However, results provided in the present thesis are far from being exhaustive and cancer cell behaviour on 3D electrospun scaffolds is worth of further thorough investigation.